Author of the award-winning AL[...]

STEVE JONES

'Jones is the Alan Bennett of science writing'
Financial Times

CORAL

A PESSIMIST IN PARADISE

Steve Jones is Professor of Genetics ay University College London. He has been awarded the Royal Society Michael Faraday Prize for science communication. He delivered the BBC Reith Lectures in 1991, appears frequently on radio and television and is a regular columnist for the *Daily Telegraph*. Previous books include *The Language of the Genes*, *Almost Like a Whale*, *Y: The Descent of Men* and *The Single Helix*.

'[Jones's] speciality is demystifying the mysteries, binning the jargon, putting the work done in labs into the lives of people . . . In its eloquence, its soundness of judgement, the vigour of the writing, and, above all, the way in which he distils the complexities unravelled by others [*Coral*] achieves a high distinction all of its own' *Sunday Telegraph*

'Jones rallies literature, politics, myth and commitment to the cause of preserving coral reefs . . . His point, wittily, pithily and passionately put, is that the destruction and exploitation of coral will inevitably bring about our own extinction' *The Times*

'A welcome addition to the canon of modern scientific monographs . . . Jones stays lightly informative throughout, and maintains a nimble poise between fact and human interest . . . In its easy erudition, its scope, its simplicity and clear humanity, *Coral* echoes the achievements of the seventeenth-century grandfather of the genre, Sir Thomas Browne' *Telegraph*

'A hugely accessible account . . . A seamless fusion of science, history, literature, politics and myth by a passionate advocate for the environment' *Herald*

'Jones is an experienced, poetic pilot, steering an eccentric and surprising course through the intricacies of reef life and its implications' *The Economist*

'Jones writes beautifully . . . *Coral* is both deeply original and a fascinating (if frightening) read' *Times Literary Supplement*

'One of science's best writers' *Guardian*

Also by Steve Jones

THE LANGUAGE OF THE GENES

IN THE BLOOD

ALMOST LIKE A WHALE

Y: THE DESCENT OF MEN

THE SINGLE HELIX

CORAL

A Pessimist in Paradise

Steve Jones

ABACUS

First published in Great Britain in 2007 by Little, Brown
This paperback edition published in 2008 by Abacus

A CIP catalogue record for this book is available
from the British Library.

ISBN 978-0-349-12147-5

Typeset in Bembo by M Rules
Printed and bound in Great Britain by
Clays Ltd, St Ives plc

Abacus
An imprint of
Little, Brown Book Group
100 Victoria Embankment
London EC4Y 0DY

An Hachette Livre UK Company
www.hachettelivre.co.uk

www.littlebrown.co.uk

To Babette and Roland Bec – reef-dwellers

THE STRUCTURE AND DISTRIBUTION OF CORAL REEFS.

BEING THE FIRST PART OF
THE GEOLOGY OF THE VOYAGE OF THE BEAGLE,
UNDER THE COMMAND OF CAPT. FITZROY, R.N.
DURING THE YEARS 1832 TO 1836.

BY
CHARLES DARWIN, M.A., F.R.S., F.G.S.,
NATURALIST TO THE EXPEDITION.

Published with the Approval of the Lords Commissioners of
Her Majesty's Treasury.

LONDON:
SMITH, ELDER AND CO., 65, CORNHILL.
1842.

Contents

PREFACE

DAVY JONES'S ATTIC

The BBC's *Desert Island Discs* is the third-longest-running radio programme in the world. It turns on a universal fantasy: of escape to a tropical paradise, in harmony with Nature. The programme's rules allow a single book plus the Bible and Shakespeare, eight carefully chosen records and just one luxury. The idea of a sea-girt atoll as a refuge from the troubles of modern life is not new, for it goes back to *The Tempest* and has been exploited by Daniel Defoe, Herman Melville and many others.

In 1942, when the series began, the notion was understandably popular, but in the real world such places can be hell. Their inhabitants go through torment for reasons that depend more on the laws of Nature than on the power of human viciousness. A remote island can soon become a tiny but terrifying nightmare. Beautiful places make for savage societies and what appears in our limited view to be primeval calm may be but a pause in some general anarchy, above and below the waves. The risky life of a large mammal on a small patch of land often renders such a supposed Utopia both violent and unstable.

Even so, the universal obsession with a sun-drenched Garden of Eden lives on. Remote, idyllic and peopled by savages (noble or otherwise), such places enter public consciousness every day; in cartoons, on *Desert Island Discs* – my own decorative but useless object when invited onto the programme two decades ago was the stuffed body of Mrs Thatcher's then Secretary of State for Education, Kenneth Clarke – and as icons of the perfect holiday. From the rich and bored on Mustique to the hopeful poor in the Coral Casino, the reefs still cast their spell.

A century and a half ago desert islands played a larger part in intellectual life, for coral atolls were the spark that ignited the theory of evolution and altered man's view of himself. They gave Charles Darwin, on his voyage across the Pacific aboard HMS *Beagle* in the 1830s, the seed of a great idea. On that famous trip he found proof of the grandest of what he called his 'grand facts': that the Earth was not static but was filled with change. Three years after the appearance of his popular work *The Voyage of the Beagle*, the notion led Darwin to write his first scientific book, *The Structure and Distribution of Coral Reefs*, published in 1842. Seventeen years later its insight into the transforming power of time gave birth to *The Origin of Species* itself.

In his coral treatise he set out 'to describe from my own observation and the works of others, the principal kinds of coral-reefs, more especially those occurring in the open ocean, and to explain the origin of their peculiar forms,' and succeeded. Darwin's theory of how such places emerge from the depths is still accepted (even if, as is the nature of biology, life has become more complicated since his day) but the reefs tell the tale of the Earth and its inhabitants in many other ways. Some were known to the eminent naturalist, but most are new. They draw on sciences as different as genetics, chemistry, physics and astronomy. Here I try to use them as the foundation for a literary edifice written in homage to Charles Darwin's magnificent work of long ago.

My own experiences of the coral realm are less dramatic than those of my predecessor but gave rise in their modest way to this book. They began with a brooch, a brown stone in a silver setting. In age and ugliness it matched the cluttered house in West Wales in which I spent much of my childhood. My father's father, and his father and grandfather too, had all been sea-captains. They each retired to New Quay, a small Cardiganshire town filled, in my youth, with superannuated sailors. On summer days dozens of mariners – most of them known after their ships, so few are the surnames of Wales – leaned on a wall above the harbour to tell and retell the stories of their lives. Below them swung fishing boats named after the vessels – the *Lady Cairns*, the *Carnedd Llewelyn*, the *Camellia* – that had for a century and more taken the men of New Quay across the globe.

My grandfather, Captain David Owen Jones, crossed the Indian Ocean many times and brought the brooch home to prove it. Just before the Second World War he took early retirement from Shell Oil and passed thirty placid years in improving his game of bowls. On Sundays he did the same for his soul, with several visits to the strictest of the town's many Presbyterian chapels.

To the dismay of its matelots, New Quay one day became famous. In the mid-1940s Dylan Thomas moved there (as he wrote to a friend: 'Though it is lovely here, I am not'). He drank in the Black Lion, a pub a hundred yards from my grandparents' house. His companions in alcohol included seafarers, fishermen and even the odd Oxford graduate. Thomas set *Under Milk Wood*, his play for voices, in the place. Its 'hunched, courters'-and-rabbits' wood limping invisible down to the sloeblack, slow, black, crow-black, fishingboat-bobbing sea' is a precise description of New Quay. Many of the locals are to be found in its pages; and Captain Cat drank, in a variety of guises, in the Black Lion. My father knew (and disliked) Dylan Thomas, but his own paterfamilias had none of Captain Cat's romantic character, which was based on another ex-mariner a few doors down. I was convinced, however, from the moment I heard the play, that my grandmother had inspired Mrs Ogmore-Pritchard, the spick-and-span lady who would not allow the sun in before he had wiped his shoes.

Her brooch was made of coral. That Victorian object lies before me as I write. It was on display in the front room among other keepsakes: an ebony elephant with ivory tusks, a brass temple bell, a dragon-topped box and a woodcut of junks on the China Sea. I have them all. Most are no more than reminders of a time long gone, but the brooch marked the first of many unexpected appearances of its raw material throughout my life.

Because the front room was Ogmore-Pritchard territory, its owner alert for any deviation from correct behaviour, Dylan Thomas had a large but indirect influence on my future. A boy did not fit, in my grandmother's scheme of things, in the company of ornaments. As a result I spent much of my time Where I Would Not Be a Nuisance, up in the attic. There I devoured book after book. On his long and dull voyages the Captain had, for lack of other amusement, been forced to read. The fossils of his Edwardian taste had lain untouched since he returned to land and were stored in well-tied boxes (my own inability to replicate the knots was often a cause of rebuke). Some of the contents of Davy Jones's many lockers were technical – Lubbock's *Last of the Windjammers*, Ashley's *Book of Knots and Splices*, *The Admiralty Manual of Seamanship* – but they included plenty of fiction, all with a maritime flavour. Conrad was a bit dense for my eleven-year-old tastes but there were lots of lesser tales to sample.

Most are forgotten and rightly so, for who now bothers with W. W. Jacobs and his jovial cockney crewmen? Just one, a bestseller in my great-grandfather's schooldays – and he went to sea when he was fifteen – has retained much of a reputation. It impressed me even then, perhaps because of its hidden homoeroticism (although I did not know the word and barely recognised the sensation). R. M. Ballantyne's *The Coral Island* was published in 1858, twelve months before *The Origin of Species*. It sold two hundred thousand copies to an audience well primed for its Christian symbolism and keen to hear tales of British pluck in the South Seas. Its intrepid blond boys hoped to bring enlightenment to the local savages. Instead, they saw enough brutality in the world of the tropics to give nightmares to generations of their fellows.

For Ballantyne, the Cannibal Isles were an image of mankind at his selfish and violent worst. He was not far from the truth, for in his day anthropophagy was still widespread (and in 2003, a group of Fijians made a belated apology to the descendants of an ecclesiastical hors d'oeuvre of more than a century earlier).

Cannibalism was in the end suppressed and *The Coral Island*'s vision of the South Seas gave way to a less strenuous view. The atolls became picturesque. The Swiss Family Robinson – their tale also to be found in Captain Jones's sea chests – had a wonderful time on their improbable isle of New Switzerland with its lions and kangaroos. Even Herman Melville, in his first book *Typee*, painted a picture of an idyllic life in what is now French Polynesia. When bland fiction lost its charm I could always dip into a series of colonial memoirs that painted a grandiose portrait of a pattern of islands filled with loyal subjects of the Empire, happy with their simple lot.

At about the time I was sharing an attic with cannibals, William Golding was at work on *Lord of the Flies* (and thanks to an excellent English teacher I read the novel not long after its publication in 1954). His book transformed the image of the tropical oceans for ever. It was modelled on Ballantyne and his saccharine successors but had quite a different agenda. Paradise was not to be regained by the forces of progress, but was lost, never to return. In the novel's post-nuclear future a group of marooned boys revert to savagery, to be rescued from their folly by an emissary from a society that has almost destroyed itself with atomic war.

For Golding, man's basest instincts triumph in the end and life on an island colonised by children is destroyed by an inevitable struggle for existence. His view of human nature descends directly from *The Origin of Species* and its notion of universal strife.

Many others have seen a message of despair in that work's modest pages. Darwin's notion of biological conflict as the engine of change has been picked up by novelists, by philosophers and by ideologues of almost every flavour. *Coral Reefs* has escaped that fate. Even so, its humble subjects have a real relevance to human affairs, not just for their biology, alluring as it is, but as a microcosm of a great shift in world-view. Attitudes to Nature and to our place in it were transformed in the century that followed the *Beagle* voyage and have altered almost as much since the days of William Golding. The confident nineteenth-century idea of existence as progress was replaced with an uneasy sense that, in science as in society, the past is not as simple as it seemed and the future less certain than we might like. Such vague disquiet has now been succeeded by a sense of real alarm; that disaster is almost upon us. As a result, scientists, like novelists, have become pessimists in what once appeared to be an earthly paradise.

Coral brooch and *Coral Island*, Calvin and colonialism; each now has an antique air as little more than a monument to the Victorian imagination. In my schooldays I saw all of them as warnings of the dangers both of Christianity and of a life at sea. I turned instead to science, mixed with light agnosticism.

In an otherwise drab introduction to that sphere, biology stood out and corals soon made their presence felt. After the chance, seized by all schoolboys, to admire sperm down the microscope we moved on to the lower animals. My attention was one day caught by a *Hydra* – a freshwater version of the creatures that make reefs – as it writhed green and alien beneath a cloudy lens. I chose a career.

For thirty years I pursued it with more energy than success. Science is a broad church full of narrow minds, trained to know ever more about even less. In my student days in Edinburgh I spent an inordinate amount of time on the science of polyps, as *Hydra* and its marine relatives are called. Why that obscure group should be important we did not, in the obedient fashion of those days, ask; but they too were relics from the past. In 1872 Charles Wyville Thomson, then Professor of Natural History at the university, had embarked on a four-year voyage aboard HMS *Challenger*. It was the foundation of the science of the sea and the basis of a superb collection of jellyfish and corals that was, through simple inertia, still used in teaching almost a century later. As a result, my first student essay was on the biology of atolls and I visited my first great library, the National Library of Scotland, to do the research. The standard book on the subject was by a C. M. Yonge (himself an Edinburgh graduate) who, with a group of young companions, had set up a marine laboratory on the Great Barrier Reef in the 1930s. I met him long afterwards when he retired to his native city, but by then his work appeared no more than quaint. To an undergraduate of the 1960s his favourite organisms looked dull when compared to what else was going on. Molecular biology was in its infancy and great discoveries were there to be made.

Many of my contemporaries became modestly famous for their efforts in that field – and one even won a Nobel Prize – but, almost by accident, my own attentions became directed towards the genetics of snails. My tutor had worked on the dozens of unique kinds then found on the island of Moorea, off Tahiti (the model for James Michener's Bali Hai in his *Tales of the South Pacific*), and, with a certain lack of initiative, I followed in his footsteps. That was a mistake. As the years went by I gained more of a suntan than a scientific reputation until at last I was, in a match to my grandfather's crown-green bowling, reduced to writing popular books about science.

My first snail sample was collected not on a tropic isle but on a sunless Scottish dune. An introduction to Bali Hai had to wait and came at second hand. Two decades ago, my research group at University College London set up a project at Wytham Wood, Oxford University's magnificent thousand-acre field site. Several times a week we drove from our dismal 1960s block next to Euston Station, along the canyon of the Marylebone Road and through the London suburbs. Within an hour or so my colleagues and I were in a meadow surrounded by woodland, on the slopes of a green hill, with views down to a medieval village. There, in snow and in summer heat (the latter aggravated by the miasma of self-congratulation that hangs over Oxford itself) we built a snail ranch, a hundred wire cages each a metre across in which we set up a variety of eccentric experiments on those agreeable animals.

As so often in research, perspiration outweighed inspiration for it took months to make the cages and almost as long to dig them in. Just beneath the surface the soil gave way to tough white rock. Closer inspection of the flakes that flew from our pickaxes showed that it contained – it was – millions of fossils; and not of land animals, but of coral.

The Oxford Rag is the remains of a reef that once covered much of England. The stone was laid down a hundred and fifty million years ago, in the Jurassic era, when the village of Wytham was beneath a tropical sea. The formation is cousin to many other ancient structures, whose immense skeletons make up some of the globe's tallest peaks. They tell of a past when dinosaurs roamed the land and when the continents had a shape quite different from today.

When Wytham Hill was alive, corals filled ten times as much of the globe as they do in modern times. Reefs are often called the rainforests of the seas but nowadays they extend to a mere half-million square kilometres – an area just twice that of the British Isles – compared to twenty times that figure for tropical forest. Reduced as they might be, they are still an important part of the empire of life; such places are biology in miniature, a microcosm of existence on the edge. In a world altered by man, many have paid a heavy price for their risky careers.

My first experience of living coral took place, as a result, on concrete. A largely artificial islet, graced by planted palms and surrounded by a reef moved in part from elsewhere, it lay a few miles from Honolulu. Coconut Island had been used as a private amusement park and as a set for a dire TV show called *Gilligan's Island*. It was visited by, among others, Presidents Truman, Johnson, Nixon and Reagan and is now a University of Hawaii marine laboratory, many of whose scientists devote their lives to studying the decline of the oceans. My baptismal trip to a Darwinian Shangri-La was an ersatz and somewhat melancholy event.

The creatures that helped build the shores of Coconut Island have become the raw material of dispute in many parts of biology. Darwin could never have guessed where his work would lead. *Coral: A Pessimist in Paradise* tries to unite a series of what seem disparate sciences in the context of the reefs, from their adventurous past to their uncertain future. It uses them as a narrative device as much as a subject, and is as a result a broader (and notably shallower) version of *The Structure and Distribution of Coral Reefs* itself.

A journey to an unfamiliar landscape introduced Charles Darwin to the notion that flesh might be as flexible as is rock. Although my own book remains unmarked by any element of originality, it emerged – like his – from many months spent in a territory that bears witness to the forces that made the atolls. I am a member of that league of British authors who have set up shop in the land of Lamarck. Here, on the slopes of the Massif Central, in southern France, the house in which I write sits between ancient ocean floor and volcano.

A few kilometres to the east is a spectacular mountain of tumbled rocks made of a magnesium-rich limestone called dolomite. The Montagne de Seranne, as it is called, was once – like Wytham Hill – a coral island in a tropical sea, but the stone has been so extensively altered that its nature was for many years disputed. The problem was solved by Marie Stopes, a geologist at my own workplace, who is perhaps better known for her interest in married love. Many marine beds are full of fossils. Dolomite is not, for its parent material was smashed by breakers and by animals that bored through its structure. The rock was ground into sand, compressed and chemically altered. In the years since the ocean fell back, the

marine remains of Seranne and it many fellows have been sculpted by the rain into monoliths, visited by thousands of tourists each year.

Other nearby geological excitements have a resonance in the reefs. A spire of basalt an hour's walk from my front door was squeezed from the depths when Africa hit Europe and built the Pyrenees. On it perches the ruins of a thirteenth-century fort, the Château de Malavieille, its history almost forgotten. The black rock from which it is made – rare on land and a frequent target of quarry-owners in search of raw material – covers more than half the Earth's surface, for it forms most of the sea floor. Liquid basalt bubbles out of the volcanic cracks that make the mid-ocean ridges and, as it spreads, moves continents around the globe.

The spike of Malavieille has a close tie with Darwin's voyage, for the coral islands of the Pacific are – as he was the first to realise – each built on a solid peak of basalt extruded from below. As the animals do their work around its shores, the mass of rock sinks under its own weight back into the Earth's crust, to leave an atoll – a white stone circle with a central lagoon – afloat in the ocean like a buoy upon a wreck.

Darwin's theory was much criticised and remained unproven for more than a century. The crucial piece of evidence had an unexpected tie with the Languedoc's geological equation. The forests that covered the steamy swamps of prehistoric France rotted away and acidified the local rivers. In turn, they etched the rock of a range of granite peaks and dissolved the uranium salts held within. The heavy metal was laid down as ore. A mine not far from Malavieille used that sinister substance to produce the fuel for nuclear power stations – and for the bombs that France once tested on a colonial atoll. In the 1960s, after public outrage put a stop to the atomic powers' aerial tests, drills were pushed deep into coral islands to sink the weapons to their explosion site. There the geologists found final proof of the notion of each atoll as a limestone chateau perched on the summit of a doomed basalt peak. The Cold War upheld Darwin's audacious claim about the plastic nature of the Earth.

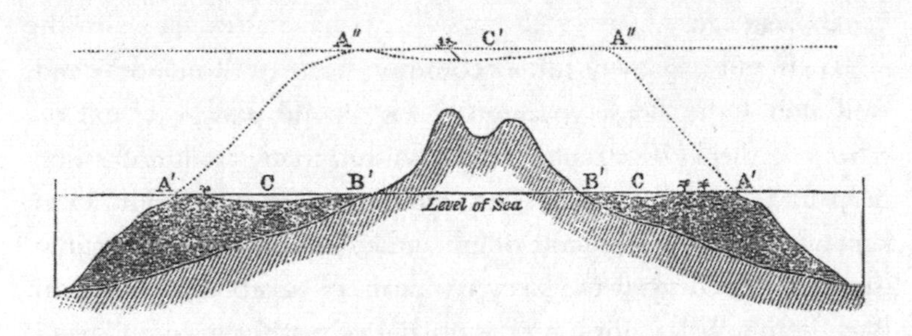

A′A′ Outer edges of the barrier reef at the level of the sea. The cocoa-nut trees represent coral-islets formed on the reef.
C C —The lagoon-channel.
B′ B′—The shores of the island, generally formed of low alluvial land and of coral detritus from the lagoon-channel.
A″ A″—The outer edges of the reef now forming an atoll.
C′—The lagoon of the newly-formed atoll. According to the scale, the depth of the lagoon and of the lagoon-channel is exaggerated

The record of the reefs has tested his ideas in many other ways. It sheds light on subjects that at first sight appear far removed from the tropical seas. They point at life's great disputes: between rising seas and falling landscapes, between order and disorder and between the survival of genes as against the death of those who bear them. From continental drift to conservation, from stem cells to sexual disease and from gerontology to the greenhouse effect, the stone from which I take my title gives an insight into some unexpected parts of science.

Diverse as they might seem, those topics share a theme, for in their various ways they reveal that Elysium has a darker side. In the century and a half since Darwin, the notion of Nature as an evolved and rational machine has been superseded by an uneasy sense that what seems like peace is filled with strife and that disorder can break out in many ways. The era that gave the proof of Darwin's ideas was one of tension between optimism and despair. This book tells the tale of how, on desert islands, that antagonism began long ago.

Reefs and negativity found common cause well before – and well after – the *Beagle* voyage. On his second passage of exploration, in the 1770s, Captain Cook visited many atolls and sailed deep into Antarctic waters. The ship's astronomer William Wales kept a log, in which he wrote of his own low spirits (almost certainly the result of scurvy, with its key symptom of 'scorbutic nostalgia'). Back home, Wales took a post as teacher of mathematics at Christ's Hospital in London. His pupils included the young Samuel Taylor Coleridge. There, no doubt, was born the Ancient Mariner, the albatross of guilt and the weary, weary times which still mark the literary, social and scientific image of the coral empire.

Many other travellers have had their illusions shattered by the truths of the tropics. The French navigator Louis-Antoine de Bougainville was the first European to land in Tahiti, in 1768. He was enraptured by the place. Sex was its main attraction. He named the island Nouvelle Cythère, after the Greek island where once had flourished the cult of Aphrodite. As he wrote on his departure: 'Farewell, happy and wise people; remain always as you are now. I will always remember you with delight and as long as I live I will celebrate the happy island of Cythera: it is the true Utopia.' His reports reached the ears of Rousseau, who in response developed his idea of the noble savage, happy in his isolation from the modern world.

Rousseau's idea is, of course, Utopian. Coral islands are not always happy places inhabited by wise people. They are often wrecked by disaster and those who live upon them can be just as volatile. The anthropologist Margaret Mead painted 1920s Samoa as an idyll of sexual equality, but overlooked the fact that rape was more frequent than in the United States. Micronesia now has more youth suicide than anywhere else, with one young man in forty choosing self-destruction. Optimists tend to ignore such problems; even Bikini, the site of the first post-war atom tests and today battered by nuclear explosions with its people driven into exile, is now sold in travel magazines as a mildly radioactive Arcadia, the ideal destination for the intrepid tourist.

De Bougainville's contented Tahitians were in truth involved in a bitter religious war which called for human sacrifice. New Cythera lost more of its shine when, soon afterwards, Captain Cook and his men stepped ashore and enlisted its inhabitants into the contaminated continent of humankind. The Tahitians referred to syphilis as the British Disease but were diplomatic enough to insist that it had been introduced by de Bougainville. Whoever was to blame, that infection was an introduction to a whole universe of illness. Only now are the peoples of the far Pacific on their way back towards their numbers before civilisation arrived.

The animals that built Nouvelle Cythère and its fellow islands have lives as unsettled as those of their human inhabitants. Again and again they show the ambiguity of the atolls. Many polyps are immortal, for they renew themselves almost without limit from special cells identified two centuries ago in *Hydra* as 'sleeping embryos'. Stem cells, as they are now called, are sometimes touted as miracle cures, but if they awake before their time they may cause cancer. The illness is an outbreak of selfishness in the body's coalition of cells. Such behaviour is universal in the tropical shallows. A year before Darwin's death a certain sea-anemone was found to live in 'reciprocal accommodation' with an alga. From that grew the science of symbiosis, the interaction of different species in what looks like altruism. The coral symbiosis is the best understood of all, but what once looked like harmony is no more than stalemate in a marriage on the edge of divorce.

To Darwin, atolls showed how life could thrive in difficult conditions. He wrote of their builders with admiration: 'We feel surprise when travellers tell us of the vast dimensions of the Pyramids and other great ruins, but how utterly insignificant are the greatest of these, when compared to these mountains of stone accumulated by the agency of various minute and tender animals!' Since his day, their image has undergone a subtle change, for those emblems of biological fortitude have become a canary in the ecological coal mine. Throughout the history of life, desert islands have always been among the first places to collapse in stressful times and they are doing the same today. The unique Moorean snails, the familiars of my student days, are gone, victims of an idiot 1970s attempt at biological control in which predatory molluscs were brought in to eat introduced African land snails. Unfortunately the predator preferred the natives (a tiny remnant is on show in London Zoo, to general indifference). Their fate is a microcosm of a greater disaster.

Today's decline of the reefs is a mirror of a catastrophe of long ago. Once again, the evidence is written in the landscape around my own retreat. The stormy geology that built the basalt point of Malavieille gives local villages a gloomy air to match that of their inhabitants. Many devote their lives to complaints about English incomers, but their ancestors had more to grumble about, for they were coal miners.

Their raw material was laid down in ancient swamps. The industry's sole relic is an enormous miner's lamp on a nearby roundabout, but in its day it did its bit to add carbon to the air and to set the stage for the greenhouse effect. Coral is an alliance of carbon with calcium and oxygen – and so are all plants and animals, ourselves included. Carbon we all are and unto carbon we shall return. With every breath and every death we take part in a series of transactions in which the currency of life moves through the reserves held in the soil, the skies and – most of all – the seas. Its slow revolutions made the modern world, but that world has led to a vast acceleration in the rate at which the machinery rotates. It may soon break down, with disastrous consequences for the corals and for ourselves.

The limestones of the Seranne Mountain formed a small part of an ancient ocean's vast reservoir of the element. The reef that made them died in a carbon crisis, the greatest disaster ever to hit our planet. Parts of the landscape nearby glow red in the sun because the soil contains iron laid down when France was in the tropics. In many places that crimson stone is capped by a sheet of sterile grey sediment. The rock was made at the end of the Permian era as a sudden burst of carbon dioxide spewed out of the ground, changing the climate and killing almost everything upon it. The tourists who flock to the area are marvelling at a geological graveyard.

Many of today's travellers in the tropics have the same experience. Next to my grandmother's brooch my desk now bears fragments of coral gathered a few months ago on a trip around the dead reefs of the Seychelles. As we sailed from island to island we were pursued, or so it felt, by what looked like a substantial cruise ship. The yacht *Pelorus* belongs to Roman Abramovich, who became a billionaire after the privatisation of Russia's oil. His return to the equator from his native wastes is a reminder of how much energy the rich parts of the world use to maintain a tropical climate and of the reckless way in which it has been dissipated. As I wrote the first paragraphs of this book during the French *canicule* of 2003, which killed thousands, I had to wipe sweat off my keyboard; I finished it three years later as Britain experienced its hottest month since records began. At the turn of the millennium the Seychelles, too, suffered a severe heatwave. More than nine-tenths of all its corals were lost. A 2005 repeat of an ecological survey made ten years earlier showed that a once complex ecosystem had been reduced to seaweed-covered rubble.

The Seychelles disaster is marching across the globe. At the time I read *Lord of the Flies* about a third of the Great Barrier Reef was still in good health. When I first went to the place a decade ago, just a quarter was in reasonable shape; now the figure is one part in five. Life has seen five major extinctions since it began. The reefs have been witnesses to them all and are now horrified onlookers to the sixth. They remind us of our own fragility and of how a Garden of Eden can so easily be destroyed. Those who live upon such places, or study them, are right to feel a certain sense of gloom.

As is the habit of the academic hack, I have lectured – with more or less conviction – on most of the subjects illuminated by the polyps. Many of my talks were given in a place with a strong tie to their history. I begin my introductory biology course each year with the statement that I am speaking from Darwin's bunker – which is true, for the basement lecture theatre at University College London is on the site of his London home, in Bloomsbury (a blue plaque and a dismal post-war block now mark the spot). The students look blank.

Charles Darwin moved to 'Macaw Cottage', as he called the house after its colour scheme (his wife insisted that the dead dog in the garden be removed first), in 1839 and stayed for just three years. He could not stand the clamour of Gower Street, which still roars with traffic (although the tyres of the Number 73 bus are quieter than the iron wheels of a hansom cab). He soon moved to the outer suburbs, where he spent the rest of his life.

In spite of the noise the young naturalist got a lot done and he wrote of his days in Bloomsbury that 'the greater part of the time, when I could do anything, was devoted to my work on "Coral Reefs", which I had begun before my marriage . . . It was thought highly of by scientific men and the theory therein given is, I think, now well established.'

Darwin was right about his theory, which has developed in directions he could never have guessed, but his account of his time in Macaw Cottage includes a telling phrase, his first reference of many to a 'frequently recurring unwellness' which interfered with his ability to write and in the end led him to become almost a recluse. Thousands of words have been spent on what that illness might have been. Some point at a psychosomatic disorder brought on by a clash between Darwin's science and the society in which he lived, while others insist that the culprit was a tropical parasite. Whatever the truth, the book Charles Darwin produced as the symptoms of a life-long malaise first showed themselves led directly to his great insights into time, change and evolution.

In his autobiography Darwin records that even in his healthier days 'Milton's "Paradise Lost" had been my chief favourite, and in my excursions during the voyage of the "Beagle", when I could take only a single volume, I always chose Milton.' Joseph Banks, botanist on the first Cook expedition, discovered hundreds of new plants on the journey. The specimens were flattened between sheets of paper, and they too came from remaindered sheets of a failed edition of Darwin's favourite reading. The choice was, we can see with hindsight, appropriate indeed, for today's science of the atolls gives that poetic theme an ever-greater resonance.

My own book is based on a global shift from hope to despond. It has made me, a land-bound scion of ancient mariners, into a sadder and (possibly) a wiser man for it shows how, in the three centuries since Europe's entry into the world of the corals, a paradise has been lost, and asks why.

As scientists make science in a way that historians do not make history I have tried in addition to tell the story of a few of the eccentric characters who grace the biology, the geology and the chemistry of the universe of the reefs. I have been helped by the generous comments of many people, several far more expert on the subjects covered than am I. I thank in particular Robert Cowie, Stephen Guise, Nigel Marsh, Shauna Murray, Norma Percy, Peter Robinson and Rachel Wood. The illustrations come directly from Darwin's own book.

The *Desert Island Discs* theme tune, with its added seagulls and waves, is almost painfully familiar to Britons of a certain age. Its title, 'By the Sleepy Lagoon', sums up the old, hopeful and simplistic view of the universe of the corals. Now, life has become more complicated and less cheerful, but the tale of those animals has emerged as a thread that binds much of biology together. The new millennium, with its concerns about the future, gives a great

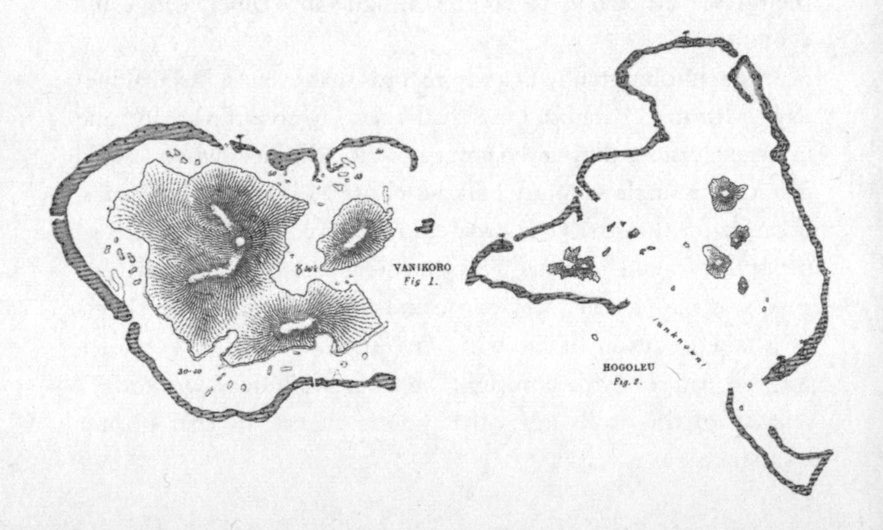

chance – perhaps the last – to find out how they work. *Coral: A Pessimist in Paradise* aims to show how their efforts have formed the past and the present of our planet, and how their plight warns us that unless we mend our ways our own future is gloomy indeed.

CHAPTER I

THE KING OF COCOS–KEELING

In 1891 a French vendor of non-perishable tarpaulins arrived in Tahiti. His ticket was subsidised by his native land (then the colonial power) and he had been enticed to the island by an advertising brochure: 'Born beneath a sky that sees no winter, on a soil of wondrous fertility, the Tahitian has only to raise his arm to pluck the fruit of the breadfruit tree and the plantain, which form his staple foods. So he never works and fishing, whereby he varies his diet, is for him a keen source of pleasure.' That image originated in the novels of Pierre Loti, a bisexual matelot whose erotic but costive works inspired van Gogh, Dumas *fils*, Mallarmé and, above all, Paul Gauguin, the tarpaulin salesman whose paintings became the icon of the South Seas.

Soon after his arrival Gauguin took up with the teenager Tehamana, the first of his many native mistresses. After a career in which his artistic talents remained almost unrecognised, he moved to the Marquesas Islands where at the age of fifty-four he died drunk, syphilitic and addicted to morphine. His shrine – a figure half-fetus, half-Diana – stands close to his grave on those islands, a thousand miles to the north and east of where he first landed.

Nowadays, Tahiti makes a great fuss of its national artist. The isle is full of postcards of his famous maidens, clad in floral necklaces and little else. Gauguin and Tahiti are as inseparable as are

Darwin and the Galapagos and it is impossible to imagine either place without at once conjuring up its most famous visitor.

An earlier traveller to the island is less fêted by posterity. Of his base at Point Venus nothing is left except a black sand beach, a wave-fretted reef and a lighthouse. The monument that graced the spot has been destroyed and tourists have their attention fixed instead on their own version of Shangri-La, which involves, in homage to its syphilitic begetter, controlled nudity bedecked by flowers.

The lost marker at Point Venus was not to art, but to science. On that spot, on his first voyage of discovery in 1769, Captain James Cook set up an observatory. There he measured a physical constant that was central to our comprehension of the solar system and, long afterwards, became a first step on the road to the atom. Boswell's 'grave steady man . . . with a ballance in his mind for truth as nice as scales for weighing a guinea' was part of an international scientific project that involved France, Italy and other countries. His monument – a sphere into which was built 'Cook's Stone', a coral slab with a brass plate – met its end in the spirit of nationalism, for many of today's islanders see the explorer as no more than an invader who brought war, misery and disease. In the 1980s they pulled the monument down (which was less outrageous than it seems, for the famous Stone was in fact a relic left long afterwards by an American sailor).

Captain Cook's plan was to observe the transit of the planet Venus across the disc of the Sun. The event happens twice a century. He hoped to estimate the distance between our own planet and its local star as the first step to a measure of what the astronomer Edmond Halley, who came up with the idea, had called 'the immensities of the celestial spheres'.

The logic was simple. People at opposite ends of the globe would see Venus cross the solar disc along somewhat different tracks, just as fans at each end of a bench in a football stadium have a slightly different view of a striker's run at goal. Because the distance between the observers (be they fans or astronomers) is known, geometry can tell us how far it must be to the player, or the celestial body. As the relative positions of the planets in the solar system had already been inferred from their patterns of movement, it becomes easy to work out the distance to the Sun.

An attempt on the previous transit, eight years earlier, had failed in several ways. The British observers Mason and Dixon – later famous for their maps of North America – were attacked by French warships on the way to Sumatra, while the French astronomer Guillaume le Gentil arrived too late at his station in India for the 1761 event and decided to wait for the 1769 repeat – which, on the day, was hidden by cloud in a normally sunny season (when he got home, eleven years after his departure, his relatives had given him up for dead).

The Royal Society was determined not to relive that experience and so sent Cook, equipped with the latest instruments, to a more or less guaranteed clear sky.

The important moments were when the black dot of Venus entered and left the face of the Sun. Cook feared that he had failed. In his diary for 3 June he noted that 'we very distinctly saw an Atmosphere or dusky shade round the body of the Planet which very much disturbed the times of the Contacts.' His doubts were not justified, for later estimates show that his figures were within a single percentage point of the correct value of the distance to the Sun as measured with radar, which is just 149,597,870 kilometres.

Cook's aristocratic passenger, the botanist Joseph Banks, had more success on the day. He showed 'three handsome young women' the image of Venus as she crossed the Sun and, as Apollo's chariot sank in the west, managed to persuade them all to sleep in his tent ('a proof of confidence which I have not before met with upon so short an acquaintance'). He was not alone in his venereal pursuits, for many of the *Endeavour*'s crew had formed relationships, at first based on commerce but later grown into affection. In Tahiti they saw public copulation, a religious act by the Ariori class, who specialised in sex accompanied by erotic dances and smothered any children that might result. The sailors were not concerned with such theological niceties. As one of them wrote, 'what is a sin in Europe, is only a simple innocent gratification in America', and before he could set out on the next leg of his voyage Cook was forced to take hostages to get back his deserters.

Captain Cook was himself impressed by the island and its inhabitants: 'They have all fine white Teeth, and for the most part short flat Noses and thick lips; yet their features are agreeable, and their gaite graceful, and their behavior to strangers and to each other is open, affable, and Courteous.' Six decades later, Charles Darwin landed at Point Venus. He too admired the people of the beach: 'A white man bathing by the side of a Tahitian was like a plant bleached by the gardener's art compared with a fine dark green one growing vigorously . . . the women appear to be in greater want of some becoming costume even than the men.'

Darwin, Banks and Cook's gratified crewmen set an agenda that lives on today. Now their successors return in their millions to the South Seas in search of what thrills might still be available. They flock to Captain Cook's point with its backdrop of rugged mountains. Each year hundreds of yachts call. They are part of the vast fleet of pleasure-seekers that girdles the globe with the help of diesel engines, satellite navigation and glossy magazines.

The *Endeavour*'s cruise was not quite as easy. After the Transit, Cook opened his sealed orders and discovered that his next task was to 'put to Sea without Loss of Time . . . making Discoverys of Countries hitherto unknown . . . Whereas there is reson to imagine that a Continent or Land of great extent may be found, you are to proceed to the southward in order to make discovery of the Continent abovementioned.'

A few months after leaving the atolls, Cook reached Australia. On his way northwards along the coast, his vessel became enmeshed in the Labyrinth, as Captain Cook called the Great Barrier Reef. It was as dangerous as it looked. Soon the inevitable happened: 'A few minutes before 11 when we had 17 [fathoms] and before the man at the lead could heave another cast, the Ship Struck and stuck fast.' After a desperate struggle, in which cannons and ballast were thrown overboard (to be rediscovered by divers in 1969), the vessel was freed. For weeks Cook could find no passage to the open sea but at last, through the Providential Channel, the *Endeavour* escaped. In a few months he was home.

The expedition was as well equipped for biology as for astronomy: 'they have all sorts of machines for catching and preserving insects; all kinds of nets, trawls, drags and hooks for coral fishing, they have even a curious contrivance of a telescope, by which, put into the water, you can see the bottom at a great depth, where it is clear'. Its leader knew that the Labyrinth, like the atolls of the Pacific, had been made by organic beings but did not enquire into which ones: as Joseph Banks noted in his journal, 'I have often lamented that we had not time to make proper observations upon this curious tribe of animals, but were so intirely taken up with the more conspicuous links of the chain of creation as fish, Plants, Birds &c &c. that it was impossible.'

The Captain noted also that the Barrier soared out of the lifeless deeps. It was 'a wall of Coral Rock rising all most perpendicular out of the unfathomable Ocean . . . the large waves of the vast Ocean meeting with so sudden a resistance making a most terrible surf breaking mountains high'. On his second voyage, two years later, he saw that in South America, far inland, lay the remnants of corals. How did reefs spring from such depths, or rise to such a height?

Cook's question marked the beginning of a great shift in scientific thought. Why, if our planet is the work of a divine architect, fixed in a perpetual mould, is the present so different from the past? Why are marine fossils found on land? How can remote islands made by sun-loving creatures emerge from a deep and dark ocean?

In 1834 the twenty-five-year-old Charles Darwin, homeward bound on the *Beagle*, came up with most of the answer. He too had seen the remains of marine life high in the Andes. His explanation led to a shift from the ancient and static view of history to the modern world of change, in the Earth and all that live upon it. From his youthful observations on coral Darwin developed the theory that changed biology.

In his student days in Edinburgh Charles Darwin had hated his lectures on the science of the rocks. Under the influence of Adam Sedgwick in Cambridge he changed his views, and by the time of his first trip ashore on the *Beagle* voyage was an enthusiast: 'The geology of St. Jago is very striking, yet simple; a stream of lava formerly flowed over the bed of the sea, formed of triturated recent shells and corals . . . It then first dawned on me that I might perhaps write a book on the geology of the various countries visited and this made me thrill with delight.'

From that day on the *Beagle*'s supernumerary passenger was as much a student of the ground beneath his feet as of the life that flourished upon it. He wrote home that geology was the ideal science to begin as it required nothing but a little thinking and hammering. Soon his diary was filled with notes about stones, mountains and minerals.

As the ship made her way across the Pacific, Darwin became intrigued by its many atolls or 'lagoon islands', as he called them. As he put it in *The Voyage of the Beagle*: 'Every one must be struck with astonishment, when he first beholds one of these vast rings of coral-rock, often many leagues in diameter, here and there surmounted by a low verdant island with dazzling white shores, bathed on the outside by the foaming breakers of the ocean and in the inside surrounding a calm expanse of water, which, from reflection, is a bright but pale green colour.'

An atoll is a dead structure made by living beings. Most of its limestone mass is built by a thin veneer of flesh, no thicker than a jar of jam spread over each square metre of stone. The polyps are the main players. Those simple animals, solid relatives of the soft-bodied *Hydra* that so impressed me at school, appear at first sight to be little more than an open bag with tentacles. Ninety-nine per cent of them live in the sea, with only a few – such as *Hydra* itself – at home in fresh water. Just a thousand or so of the ten thousand species now known lay down rock. They were once thought to be primitive creatures of little interest to science. In fact, they are sophisticated beasts who belong to one of the most diverse of all living groups and play a large and increasing part in modern biology.

Their close relatives the jellyfish float in billions through the oceans. Some are tiny, others huge – longer than a blue whale, with tentacles thirty metres long. Jellyfish find their headquarters in the icy blackness of the sea, two kilometres down, where they pullulate in places where little else can live. Most of the reef-building corals, in contrast, flourish only in the topmost layer of water. Like green plants, they are powered by solar energy, and – like plants – the Sun provides most of their fuel. Evolution has given the reef-builders the ability to extract dissolved marine chemicals and turn them into stone. The animals suck the raw materials of calcium carbonate, the rock that forms their home, from the ocean in a process quite like that of the human body as it lays down bone. Sea snails, sea urchins, stony algae and tiny marine animals called foraminifera all share the architectural task, and together build a structure which grows at marvellous speed. As oceans open and close and continents drift, the reef-builders have made the calcium-rich landscape that now covers large parts of the world.

Their homes are built in a variety of styles. Atolls – small, isolated and elegant – are the lady-chapels of marine architecture. Many tropical coasts are fringed by mile after mile of rock formed in utilitarian and somewhat Norman sameness, with occasional gaps where rivers enter the sea. The Great Barrier Reef is the St Peter's of the polyp realm: huge, grandiose and filled with unexpected artefacts.

Most reefs have an outer face exposed to the seas and another in calmer water. The outer edge grows best for it has abundant oxygen and food, and the action of the waves causes the sea to lose some of its carbon dioxide, the acidity of which slows the rate at which raw material is laid down. The quieter coast – around a central lagoon or a channel close to the mainland – is less active. In the shallower water behind the ocean front new rock cannot extend upwards, so the mass becomes wider instead. If the water is clear the builders may stay alive a hundred metres beneath the surface, but in muddy coastal seas they do less well.

Joseph Banks had thought the animals that made such places to be curious indeed, and he was right. They are, we now know, a form of life quite different from all the others collected on the *Endeavour* voyage. They are simple enough, but over vast ages have built edifices that may – like the Barrier, the only biological entity visible from space – be two thousand kilometres long. The Barrier grows in rather a shallow sea, but most atolls soar from the deeps, in splendid isolation.

On the *Beagle*, Darwin turned his attention to an old mystery. The coral 'insect' (a term then used to describe a variety of small organisms) could not survive more than a few feet beneath the waves. How could its rocky products rise from the depths of a sunless ocean? The question was to form Darwin's philosophy. Its answer emerged on the remote Cocos–Keeling Island in the Indian Ocean, the site of the first great leap of his scientific imagination (and the only time he landed on an atoll). His insight revealed a close tie between the Pacific's lagoon islands and the story of life itself.

Cocos–Keeling has not much disturbed the course of history. The place was discovered in 1609 by a Captain William Keeling. It was

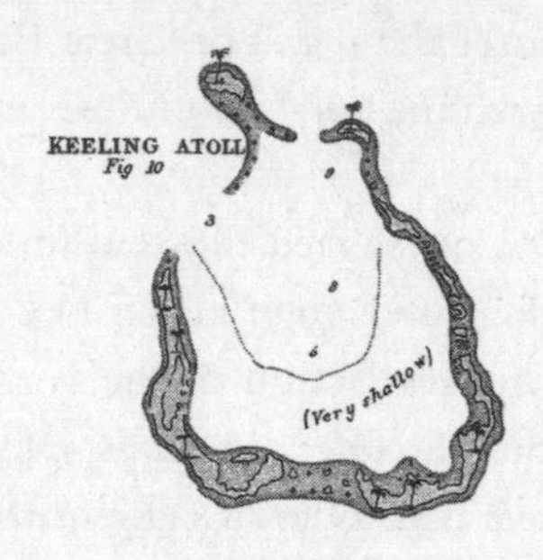

KEELING, or COCOS ATOLL, (or lagoon-island), in the Indian Ocean; from the survey by Capt. Fitzroy; scale ¼ of an inch to a mile; the lagoon south of the dotted line is very shallow, an is left almost bare at low water; the part of the north of the line is choked up with irregular reefs. The annular reef on the N. W. side is broken, and blends into a shoal sand-bank, on which the sea breaks.

forgotten until 1825, when the Scottish adventurer John Clunies-Ross – Ross Primus, King of Cocos, as he styled himself – set foot upon its empty shores. He had married an English girl who rescued him from a London press-gang and insisted that they should live in a place where the sea would always sound in his ears. Cocos–Keeling – a circle of islets a dozen square kilometres in extent, three thousand kilometres west of where the Australian city of Darwin now stands – seemed ideal. With the help of Alexander Hare (in Charles Darwin's words a 'worthless character') who arrived with a group of a hundred slaves from the Malay Peninsula, they set up a colony.

Hare insisted on his rights to a harem, his 'fiddle-faddle' as he called it. His Presbyterian partner disapproved. There were quarrels, the fiddle-faddle decamped to join the opposition, and Hare fled. In time Cocos–Keeling became a modest Utopia. Ross planted coconut groves, founded a school and wrote a treatise on the works of Malthus. Each family had a house and a garden and there was free education for all. So concerned was the Ross dynasty that its Eden might be defiled that nobody who left was allowed to return for fear of infection by new diseases, or by new ideas.

Ross's kingdom was annexed by Britain in the 1860s, under the impression that it was a different set of Cocos Islands in the Andaman group. Queen Victoria gave his dynasty the right to rule in perpetuity. Their domain gained modest fame as the site of the first battle engagement of the Royal Australian Navy and, less gloriously, of the only executions of Commonwealth soldiers during the Second World War, when three Ceylonese were hanged for their plot to surrender the place to the Japanese. The Clunies-Rosses held sway until the 1950s, when they were deposed by the government of Australia, with considerable bitterness, a local insurrection and – for the quasi-royal family – poverty and exile. Most of today's six hundred inhabitants are descendants of the first settlers and although the economy no longer depends on coconuts it is much helped by the tourists who come to see the spectacular marine life.

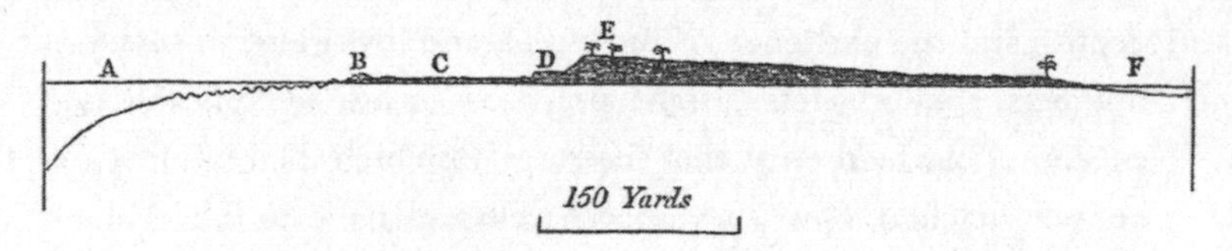

The accompanying woodcut represents a vertical section, supposed to be drawn at low water from the outer coast across one of the low islets (one being taken of average dimensions) to within the lagoon

A—Level of the sea at low water: where the letter A is placed, the depth is 25 fathoms, and the distance rather more than 150 yards from the edge of the reef.
B—Outer edge of that flat part of the reef, which dries at low water: the edge either consists of a convex mound, as represented, or of rugged points, like those a little farther seaward, beneath the water.
C—A flat of coral-rock, covered at high water.
D—A low projecting ledge of brecciated coral-rock, washed by the waves at high water.
E—A slope of loose fragments, reached by the sea only during gales: the upper part, which is from six to twelve feet high, is clothed with vegetation. The surface of the islet gently slopes to the lagoon.
F—Level of the lagoon at low-water.

The place of Cocos–Keeling in science is secure, for Darwin used his stay to develop ideas that had come to him in his months on the coast of Chile. As he wrote in his *Autobiography*: 'No other work of mine was begun in so deductive a spirit as this; for the whole theory was thought out on the west coast of South America before I had seen a true coral reef. I had only therefore to verify and extend my views by a careful examination of living reefs.'

Lagoon islands, he suggested, were born around oceanic peaks that had been thrust up from far below and then sank. His book puts the case:

> The facts stand thus; there are many large tracts of ocean, without any high land, interspersed with reefs and islets, formed by the growth of those kinds of corals, which cannot live at great depths; and the existence of these reefs and low islets, in such numbers and at such distant points, is quite inexplicable, excepting on the theory, that the bases on which the reefs first became attached, slowly and successively sank beneath the level of the sea, whilst the corals continued to grow upwards. No positive facts are opposed to this view and some general considerations render it probable.

First, the mass grew around the shores of a newborn volcanic island to generate a fringe of rock, like those Darwin had seen in Tahiti and elsewhere. The huge lump that formed its foundation vanished little by little over millions of years back into the planet's interior and was at the same time eroded by the rain and the waves. The animals continued to grow upwards. For a time, a spike of native rock might be left in the centre of a remote lagoon, but eventually the marine mountain disappeared to leave a circle of white stone in the distant ocean. On Cocos–Keeling he saw direct evidence of such subsidence, for the ruins of a shed washed by the tides had, the locals told him, been above high water mark just seven years earlier. Darwin, a poor swimmer, used a leaping-pole to reach the outer margins of the atoll. There he found the polyps hard at work, for their efforts had choked a small ship-channel within a decade. The energetic animals could, he was sure, push themselves towards the light just as fast as their mountain chose to sink.

Some branched corals can manage fifteen centimetres of rock a year – six kilograms of limestone in a square metre. A large proportion is dissolved back into the sea, or smashed by the waves and by the many animals that grind and burrow through the stone, but the new material still represents a substantial tonnage, given the globe's two hundred and fifty billion square metres of reef. In the immensity of time, such structures may grow to become several kilometres thick. In Hawaii, corals begin to develop on lava flows as soon as they enter the sea. It does not take long for a mature reef to shoot up on a sterile mass of lava. With plenty of waves and a lot of food, twenty years does the job, while in quieter places it takes half a century. Many tropical versions are far older because of the fine balance of growth and dissolution that slows their expansion.

Darwin accepted that not all the products churned out by the marine factory were built on undersea peaks. Those off a stable coastline might, he surmised, grow instead from a shallow bottom. Lagoon islands, however, were proof of the then startling idea that our planet is not solid but pliant and that even mighty mountains can plunge into the depths. The process of steady change over vast time became the central element of Darwin's later and greater theory of 'descent with modification', of evolution. His work on coral marked, as a result, the birth of the modern sciences of geology and of biology.

Charles Darwin's teacher, the geologist Charles Lyell, had assumed that atolls were no more than the crests of submarine volcanoes just below the surface whose rims had been overgrown by corals. Lyell was delighted when the young explorer told him

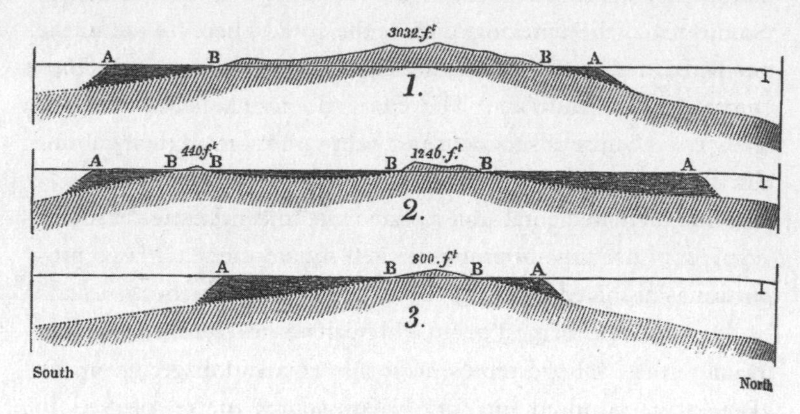

BARRIER REEFS

1—Vanikoro, from the Atlas of the voyage of the Astrolabe, *by D. D'Urville.*
2—Gambier Island, from Beechey.
3—Maurua, from the atlas of the voyage of the Coquille, *by Duperrey. The horizontal line is the level of the sea, from which on the right hand a plummet descends, representing a depth of 200 fathoms, or 1200 feet. The vertical shading shows the section of the land, and the horizontal shading that of the encircling barrier-reef; from the smallness of the scale, the lagoon-channel could not be represented.*
A A—Outer edge of the coral-reefs, where the sea breaks.
B B—The shore of the encircled islands.

of his observations on Cocos–Keeling. Darwin later wrote that the grand old man danced about, threw himself into wild contortions and conceded that he would have to abandon his own hypothesis. He saw what awaited his student: 'I could think of nothing for days after your lesson on coral reefs, but of the top of submerged continents. It is all true, but do not flatter yourself that you will be believed, till you are growing bald, like me, with hard work & vexation at the incredulity in the world.'

On 31 May 1837 Charles Darwin presented to the Geological Society in London his paper 'On Certain Areas of Elevation and

Subsidence in the Pacific and Indian Oceans, as deduced from the Study of Coral Formations'. His radical new theory was, as Lyell predicted, assailed by the establishment, and even Clunies-Ross wrote a book against it. A Dr Stanley Gardner believed that atolls grew from banks of sediment just below the surface, while others felt that such structures were a simple result of the rise and fall of the sea. The central lagoon emerged because the creatures that made coral flourished among the breakers on the seaward edge but starved or were smothered in that calm, sandy and unproductive place.

Many of the critics were powerful figures in science. James Dana, on the United States Exploring Expedition, found on the 'Feejees' reefs on an even grander scale than those seen on the *Beagle* voyage. He noted some details missed by Darwin. Most of the fine and compact stone lacked fossils – which had, he decided, been broken down by the waves and by animals that tunnelled through to rework the rock. Even so, he was dubious about the notion of vanished mountains, for the idea of a plastic Earth did not explain how they rose in the first place. Darwin's idea was, he said, 'a theory of mountains, with the origin of mountains left out'.

In December 1872 the research ship HMS *Challenger*, under the scientific direction of Professor Charles Wyville Thomson, left England on a voyage of more than a hundred thousand kilometres, with five Atlantic passages over three and a half years. The trip marked a dogged attempt to understand the ocean (as a scientist on board noted: 'It is possible even for a naturalist to get weary of deep-sea dredging'). Soon the sounders discovered great trenches several kilometres deep and a long mountain range extending south from Iceland. The high seas, it became clear, had a geological life of their own.

In the Report on the expedition, John Murray, the founder of modern oceanography, noted that the survey had revealed undersea peaks that reached to within a few fathoms of the surface. Why should coral animals not grow upwards from these, or use as a foundation the mass of minute skeletons of dead marine creatures that rained upon their surfaces, with no need for a lost mountain? Atolls were no more than a visible statement of a stable landscape of peaks and valleys beneath the waves.

The American zoologist Alexander Agassiz agreed. He launched a virulent attack on Darwin. Agassiz had studied the corals of Florida and the Caribbean and found that they sat on platforms in a shallow sea, or on points of land, with no sign of subsidence. He used the dispute to re-ignite a family feud. His father, Louis, had been Professor of Geology at Neuchâtel. Quite unawares, Agassiz Senior had himself uncovered some of the first evidence that the Earth was far from solid when he described various species of fossil fish from a bed of slate in Switzerland. They fell into two groups – short and fat, or long and thin. The thin fish were, we now know, the same as their squat equivalents, but their fossil bones had been stretched on a geological rack as the Alps slid sideways. The real number of species is half what he supposed.

Louis Agassiz was the first to see that ice ages had formed the landscape of Europe. The notion was long resisted. Darwin himself argued that the parallel roads of Glen Roy, a set of level terraces on the slopes of a Highland valley, had been formed when it was under the sea, each 'road' a beach made by waves rather than – the correct explanation – the shorelines of a series of glacial lakes at different levels.

In time he admitted that the Glen Roy episode was a failure and wrote in generous terms about the glacial theory in *The Origin*. That was not enough for the Swiss savant's son, who was a teenager when that work was published. In the tradition of his father – who had preached against evolution as 'untrue in its facts, unscientific in its methods, and mischievous in its tendency' – Alexander Agassiz became a coral contrarian. He insisted that atolls, even those in distant oceans, grew on their own rubble, with no need for a submerged peak. In time, the mass reached upwards to give a solid block that stretched from the ocean floor to the surf. After many expeditions and millions of his own dollars he was convinced that the notion of a sinking peak was 'Such a lot of twaddle – it's all wrong what Darwin said and the charts ought to have shown him he was talking nonsense.'

There was just one way to find out who was right. Charles Darwin, in the last few months of his life, wrote to the younger Agassiz with a heavy hint: 'If I am wrong, the sooner I am knocked on the head and annihilated so much the better . . . I wish that some doubly rich millionaire would take it into his head to have borings made in some of the Pacific and Indian atolls and bring home cores for slicing to a depth of 500 or 600 feet.' Agassiz was the obvious man. He was doubly rich indeed, for he owned a chunk of the world's largest copper mine and had enough spare change to pour cash into his father's famous Museum of Comparative Zoology at Harvard. At his own expense he made several boreholes on Fiji and elsewhere – but never got deep enough to reach bedrock. He stayed confident that Darwin's ideas were wide of the mark and promised a massive book in refutation but, after he died suddenly at sea, no manuscript was ever found.

In the 1890s the Royal Society sent three expeditions to the island of Funafuti in the Tuvalu archipelago in the central-western Pacific where, after various disasters with their machinery, they too found only coral and its fellows at a depth of more than three hundred metres. They had to give up when the workers revolted and the stock of diamonds for their drill-head ran out.

Science is not cheap and the many others who set off to solve the riddle of the reefs also failed, for no government would pay to bore down far enough to penetrate an atoll's rocky base, if indeed it existed.

War, too, is expensive, but there the state's generosity knows no end. Darwin's vindication came with the nuclear age. That low point in human history did a lot to destroy the romantic image of the islands, but it also showed that such places rested on solid foundations.

The power of the Hiroshima and Nagasaki atomic bombs had astonished those who made them and the post-war versions promised to be even more potent. Scientists warned of the potential of such weapons to destroy civilisation and, in a burst of euphoria in 1945 the newborn United Nations set out to outlaw them altogether.

Military minds thought otherwise. The first peacetime nuclear explosion, Operation Crossroads, was planned for July 1946. It emerged in part from rivalry between the United States Air Force and the Navy, for the admirals were stung by the suggestion that the new weapons had made their ships obsolete. Bikini Atoll, in the remote Pacific, was chosen as a test site.

The Military Governor graciously asked the local chief whether he would, as he put it, sacrifice his island for the benefit of all men. The chief responded that 'if the United States government and the scientists of the world want to use our atoll for furthering development, which with God's blessing will result in kindness and benefit to all mankind, my people will be pleased to go elsewhere'. A few weeks later 166 islanders left for the nearby atoll of Rongrik. There, they almost starved and were moved on – but they were not to return home until 1972. A new evacuation took place soon afterwards when high levels of radiation were found. In 2001 the islanders, still in exile, were provisionally awarded half a billion dollars in damages, but in 2007 lost their case for a final payout.

A battered fleet of American, German and Japanese vessels was moored in the Bikini lagoon. It carried hundreds of farm animals, the pigs dressed in anti-flash suits, the goats with their hair clipped to human length (the Southern Dairy Goat Owners' Association was outraged, on the grounds that good goats were scarcer than good Congressmen, who should be used instead). Five hundred photographers armed with almost half the globe's stock of film, together with reporters, politicians, foreign observers (and a few psychiatrists in case the event went wrong) went into action. On the day, all sat back to admire the blast.

The first test, Able, an aerial weapon that drifted off target, was rather a flop and sank just five ships. Baker, the second, submarine explosion, was more impressive: the explosion was 'so fantastic, so mighty and so beyond belief that men's emotions burst from their throats in wild shouts'. So amazed was the public by the pictures of huge vessels flung into the air that the island gave a name to a bathing costume supposed to have an equivalent effect.

Far more radiation was left in the lagoon than had been expected. Much as the General in command insisted that radiation sickness was a very pleasant way to die, his team decamped to the atoll of Enewetak and were rather more careful with their later experiments. In 1952 a test of the hydrogen bomb on that island scorched terns to death twenty kilometres away and left the lagoon's fish as charred as if they had been dropped into a hot pan. For the benefit of all men, the nuclear tests went on. Between 1946 and 1958 the equivalent of seven thousand Hiroshima bombs – more than one a day – was detonated on Bikini and Enewetak. Bikini's tropical charms had by then been transformed into a wasteland, with three of its small islets simply vaporised.

Twelve months after the Baker blast a team of scientists and engineers descended on the site. The Bikini Scientific Resurvey set out to assess the effects of Operation Crossroads. How much damage had the bomb done and how might its successors be persuaded to do still more? How far did the tremors from such explosions spread and how many of the native creatures were left? Rats did quite well and were carefully studied in the hope that they would hint at survival strategies for soldiers.

Atomic bombs, it was clear, were the weapons of the future. They posed many urgent technical questions. Cash poured in to find the answers.

The armed forces had no interest in pure science, but many of those who worked for them did. They were at first accused of 'painting their projects blue' – using Navy money to support fundamental scientific work – but as the Cold War became more and more of a threat the authorities began to see the military value of basic research. The Bikini Resurvey was one of the first moves towards the military–scientific complex that transformed our perception of our planet, desert islands included.

One prime task was to find out what lay beneath the atoll's fragile surface. The geologists had echo-sounders, seismographs and – crucially – a powerful drill. The machinery was set to work. On Enewetak it made a hole 1411 metres deep (a record at the time). After weeks in which they found nothing but coral, the bit brought forth a rock called olivine. That basaltic mineral is a tough mixture of various calcium, magnesium and iron salts combined with silicon and oxygen, which has sprung from deep within our planet to form the peaks of many of the Pacific's high islands. On Enewetak no sign of such a structure was visible on the surface, but it formed the foundation upon which the island rested. Beneath the mushroom cloud Darwin was proved right and, at once, the geologist in charge proclaimed the fact with a notice by the borehole.

Research on oceanic volcanoes has made huge steps since the Enewetak tests. Their history has been worked out in most detail in Hawaii. The chain, almost as long as the mainland United States is wide, is a microcosm of geological evolution, from vigorous young volcano to senile atoll and to drowned and distant reef. Sensors on the Big Island, the youngest of all, measure the rate of subsidence; radioactive isotopes give the age of each island, and the slope of the underwater mass compared to that above the waves shows how fast the exposed rock has been worn away. Over the seventy million years or so since the chain emerged, each new peak quickly soared to its maximum height of around a thousand metres, but then subsided over only a million years or so. As the rain beats down and gravity continues its work each peak erodes to a low mound surrounded by a white rocky fringe and, in the end, becomes an atoll. At some times there were more (and higher) islands than today, and at others there were fewer, but their history is, more or less, that proposed by Darwin. His theory gains further weight from the string of submerged atolls that stretch far beyond the visible limit of the island system and reach almost to Siberian shores.

Soon after the military expedition that proved Darwin right, Russia developed its own nuclear weapons. For the West, a series of urgent questions at once arose. Where were the tests, and how far did the fall-out spread? Politicians needed answers but, to avoid unnecessary alarm to their constituents, were obliged to hide the reason why. The International Geophysical Year of 1957–8 was trumpeted as a global partnership, a worthy successor to the efforts of Cook and Darwin but, like the voyages of the *Endeavour* and the *Beagle*, it was in the main a military exercise lightly disguised as science. Even the Russians were invited to join, but as their main contribution was to launch *Sputnik* that attempt at amity led instead to an acceleration of the arms race.

Whatever the motives, the work that began at Bikini changed our view of the reefs and of our planet. The Cold War was a great experiment in geology. Many of the underground explosions that succeeded the aerial blasts after the Test Ban Treaty of 1963 were carried out within the basalt mountains hidden far below all lagoon islands. Dozens of holes were drilled into atolls and evidence in Darwin's favour found again and again. Each test sent an echo through the globe. To pick up the reports of far-away detonations the Americans built chains of sensors across the planet. Just as in the Transit of Venus, observers in different places had a different view of each event. Geometry then made it possible to pinpoint the site of the explosion.

The pure light of the eighteenth century as it streamed down upon Point Venus had travelled in a straight line through empty space, which made it easy for the astronomers to do their sums. The nuclear shocks of the 1960s and 1970s had a more tortuous history, for they passed through the Earth before they reached the observation stations. As the figures came in and were added to earlier data from earthquakes it became clear that our planet was not a simple place. To make the sums fit together the geometers were forced to accept that the ground upon which they stood was rather more than a ball of clay.

The subterranean rumbles did not travel in a straight line. Instead, just as when light passes from air to glass, the waves changed speed and direction as they raced through the Earth. Sometimes they were reflected back to the surface and sometimes refracted onwards at a different angle, determined by the density of the material through which they passed. As the globe rang like a bell to the tolls of the atomic age the waves altered course several times. Our home, the nuclear echoes showed, has a complicated internal structure.

The planet's outer layer, the crust, a few kilometres thick, is filled with light elements such as carbon and silicon, often combined with oxygen. Below that is the denser mantle. Around the centre lies a heavy core of molten iron and various metal alloys. Within that high-pressure inferno floats a solid centre of the same stuff, two thousand kilometres across, with a mass equivalent to that of the moon and a temperature of around six thousand degrees.

Beneath our feet, the landscape is more fluid even than imagined by Darwin. In 1971, and again in 1974, a military observation post in Montana, near the President's wartime hideaway, picked up two explosions at the test site on Novaya Zemlya, in the Arctic Ocean north of Russia. The shocks made by the later bang took a tenth of a second less to arrive than had those of its predecessor. Something had changed inside the Earth. Its innermost mass is arranged in layers that pass vibrations more readily in one direction than the other. The shift in the speed of the wave showed that the central body spins somewhat quicker than does the more rigid envelope upon which we stand. In the interval between the tests it had moved with respect to the outer layer, and the shockwaves passed on faster than before. The iron lump afloat in a turbulent liquid turns the Earth into a dynamo and produces magnetic fields upon its surface. We spend our lives on the surface of a huge engine whose rotor spins once in four centuries. Its movements match the whirlpools of rock that mould the landscape far above and push spikes of basalt through the crust.

Plumes of heated rock begin in the interior and bubble their way towards the surface to form 'hot spots'. In some places the rocks may emerge as a volcano, that sinks again as it cools. The presence of large amounts of rare metals such as osmium in the lavas that built Hawaii – more abundant in the centre than near the surface – suggests to some (but not all) geologists that such plumes originate in the core itself, three thousand kilometres down. They must arise at least several hundred kilometres below the ground, for the surface plates pass over them with no disturbance. The hot spots beneath Ascension Island, the Azores, the Canaries, Samoa, Easter Island and Tahiti also seem to reach almost to the core, while those below Cocos–Keeling and the Galapagos are shallower. The best-known plume, beneath Oahu, the main island of Hawaii, is a paired complex a hundred kilometres across.

The sonic waves as they bounced also found huge lumps of dense material within the surface layer. Many still form the roots of tall peaks that reach towards the centre, while others are sunken remnants of such objects, all visible evidence of their presence gone. Even Finland, not now noted for its Alpine scenery, sits on a mass of heavy rock that was once the base of a tall mountain range.

Darwin himself believed that volcanoes could not erupt underwater; which is odd, for three months before the *Beagle*'s departure there had been extensive publicity about a new volcano that rose out of the sea between Sicily and Africa. It was claimed by Britain and France and by King Ferdinand of Sicily (who in turn named it Graham, Julia and Ferdinandea) but soon the structure sank back into the depths. In fact, such explosions of submarine fire are common. The mountain Loihi rises off the Big Island of Hawaii, its summit more than a kilometre beneath the waves. In 1996 a large disturbance made a crater three hundred metres deep on its flanks – an eruption as big as that of the 1980 explosion of Mount St Helens in Washington State. The sensors placed on its slopes pick up plenty of bangs and crackles that suggest that Loihi is very much alive. As it rumbles, it emits a sinister pall of detritus that makes a white 'snow' upon its underwater peak. At the present rate, it will break the surface within ten thousand years, to pour liquid rock down its flanks and form a new Hawaiian island.

As the Cold War dissected our home planet's interior it also improved the image of the ocean floor. Nuclear submarines were among the most dangerous weapons of that uncertain era. To skulk in watery seclusion they needed a map.

The underwater world was once assumed to be a bland sort of place. Geological operas might play themselves out on land but most of the sea bottom looked like a kind of submerged Siberia, dead, flat and dull. Many geologists doubted the very existence of more than a few marine mountains of the kind found by HMS *Challenger*, let alone the idea that they could be the foundations of hundreds of atolls.

Their scepticism came from the simple difficulty of plumbing the depths. The British Admiralty had tried, but failed. Its estimate that the Atlantic was ten miles deep depended on a line dragged sideways by currents as it was payed out. Later, twine pulled down by a massive cannonball gave a better measure. It was the first indication of the presence of a ridge in the centre of that sea. Commercial confidence was undimmed and in 1858 a cable was laid across the supposed plateau between Newfoundland and Ireland. It broke just after the celebratory banquet as a hint that the bottom of the sea was less tedious than it seemed.

Oceanography soon moved beyond cannonballs and string. The breakthrough had come well before the *Challenger* expedition. In the 1820s Jean-Daniel Colladon had tried some odd experiments with church bells submerged in Lake Geneva. An assistant hit the bell and at the same instant fired off a flare, while on the other side of the lake Colladon, armed with an ear trumpet immersed in the water, noted the delay between clang and bang. He showed that sound goes four times faster in water than in air – and founded the modern science of the oceans.

In 1900 the Canadian engineer Reginald A. Fessenden uttered the first words broadcast by radio ('Is it snowing where you are, Mr Thiessen?'). Pausing only to invent the tracer bullet and the automatic garage door opener, he turned to the use of submarine bells in navigation. After the *Titanic* tragedy the need to spot obstacles became urgent. In his system, lightships tolled a watery tocsin to be picked up by vessels far away. It was not a success.

Fessenden's iceberg-warning device used a vibrating cylinder to make pulses of underwater sound and a receiver on the ship to sense any echoes. The apparatus could pick up an object on the surface three kilometres away, but often it gave a second signal at around half that distance, for no apparent reason.

The unexpected reflection had bounced off the bottom of the sea. It made a statement of how much water lay beneath the keel. The echo sounder had been invented. At once, the US Navy seized upon the idea. In 1922 the USS *Stewart* steamed to Gibraltar and on the way made nine hundred soundings – three times as many as made by HMS *Challenger* on a voyage of three years. The Germans used the same kit on a ship that was sent to criss-cross the Atlantic in a failed venture to extract gold from the sea to pay war reparations. An outline of the ocean bottom began to emerge.

The Fessenden machine mutated into the anti-submarine device whose 'pings' became the background music of many war films. Once again, science benefited from human folly. During the Second World War, as the Battle of the Coral Sea raged, an operator who was in civil life a geologist noticed the trace of a flat-topped mountain far below. More and more such structures were found. They were named guyots, after the Swiss geologist Arnold Guyot (a man who saw in science a proof of the Bible and developed a theory of planetary evolution that began in Hades, or Africa, and ended in heaven, which he identified with the United States). An improved version of Fessenden's device sent out a beam from the bow and stern, with the reflections picked up by sensors amidships. Soon computers made it possible to paint an image of a broad strip of the bottom, a kilometre and more on either side of the survey ship.

The rays of sound transformed our perception of the ocean floor. It is not a passive and ancient basin filled with the detritus of the land, but a young and active place, an agent of change for the continents themselves. The science of plate tectonics – the study of the mobile slabs of rock upon which every continent sits – was born.

At the mid-ocean rifts, those scars in the Earth, liquid basalt gushes forth to form solid plates that grind against each other and, far away, are dragged back into the deeps. As they do, they shape our planet. Now the names of the shattered landscapes of Pangaea and Gondwanaland – the ancient continents whose remains were reassembled to make the atlas of today – are more familiar than are those of many modern nations. Three square kilometres of land are sucked into the depths each year. That may not sound a lot, but, given the immense age of our planet, means that most of the ground upon which we stand has been through its internal spin-cycle more than once. Geology hints at a history of lost super-continents, with several demolitions and reconstructions of the global map long before Pangaea.

The intestinal dramas come and go. Four hundred million years ago a lump of rock as big as the Moon slumped from an outer layer into the Earth's liquid core. That caused the globe to speed up in just the same way as a ballet dancer increases the rate of spin when she pulls her arms to her side. A blob of liquid rock was displaced to the surface and caused the eruption of vast plumes of heat. As they belched forth, the basalt built new mountains to join the others that dot the sea bottom.

To become an official sea-mount a crest must rise a thousand metres – about the height of Snowdon – above the abyss. The Atlantic's Everest is the Meteor sea-mount, which stands at more than four thousand metres above the bottom and is a hundred kilometres across. Mauna Loa on Hawaii, which rises five thousand metres out of the sea floor to the surface and a further four thousand above the waves, is the highest of all mountains when measured from base to peak.

Now such summits are tracked not with sound waves but with satellites that sense changes in gravitational field caused by the presence of a huge mass of rock. A grid has been laid from space across the entire ocean, which will soon be almost as well mapped as is the land (the job is not yet done: in 2005 a US Navy submarine ran headlong into a sea-mount as its chart was inaccurate by several hundred metres).

The life and death of a sea-mount is not like that of its equivalents on land. The Alps and the Himalayas were pushed upwards as continental plates ground against each other and like all mountains are soon washed away by the rain. Sea-mounts have a more intimate tie with the bowels of the Earth. Their birth comes as lava and they meet their end as they are battered by the waves and currents and drown in the lighter rock around them.

Their remains are swept across the globe as the continents are pushed around. Hawaii is at the southern tip of a chain of sea-mounts that stretches five thousand kilometres to the north and west. A few are still above the surface, but the desolate Kure Atoll – the northernmost lagoon island of all – is the last visible to the eye, with dozens more hidden beneath the waves between that isolated spot and the Russian mainland. The chain has an unexpected kink for about halfway to the Kamchatka Peninsula it takes a swerve to the right. The plume of hot rock rose like a slow column of smoke as the plate passed above and, long ago, changed direction. Like the mark of a candle left by an explorer on a cave roof the evidence of its track lies in the smear left behind. The hot spot itself has also moved south, as further evidence of the turmoil down below.

The Pacific has twenty thousand sea-mounts, a small minority of which bear an atoll. Whole ranges, far longer than any on land, are hidden beneath its waves. The Foundation sea-mounts, near Easter Island, stretch for more than a thousand kilometres. The Atlantic is a flatter place, with no more than a thousand or so underwater peaks. New England has a chain of its own that extends south and east from Boston.

Darwin would have loved a chance to visit a subaqueous Alp, for each range is a sort of underwater Galapagos. Their ecology is upside down when compared with that on land, for the summits are filled with life while the abyss is almost empty. The currents that beat against their cliffs put Alpine storms to shame. They bear food and oxygen which allow the upper slopes to teem with plants and animals that live on the rain of corpses that falls from the open sea above. Already such places have yielded thousands of new creatures, some from ancient groups assumed to be extinct or almost so. Many are restricted to a single peak and each chain has its own distinct flora and fauna. The thousand and more kilometres between the sea-mounts of New Caledonia and those of the Tasman Sea has produced quite a different set of inhabitants in each place. Evolution is hard at work beneath the waves.

Sea-mounts and shorelines are complicated places and so are the structures that grow upon them. Many reefs outdo jungles in their diversity. Matthew Flinders, a shipmate of Captain Bligh and an early explorer of the Great Barrier Reef, was impressed:

he saw objects that looked like 'wheat sheaves, mushrooms, stag horns, cabbage leaves and a variety of other forms, glowing under water with vivid tints of every shade betwixt green, purple, brown and white'. The struggle for existence is as pitiless and the force of natural selection as potent there as anywhere else.

A reef is a wall of mouths. Many belong to its polyps, but others are hidden away, ready to feed on whatever is on offer. The rock is riddled with holes made by animals that gnaw their way through. A large parrotfish may scrape off five tons of rock each year, transforming it into fine sand. Other borers, scrapers and grinders, from snails to fish, are hard at work. As a result, many lumps of coral resemble a Swiss cheese. They are – like our own guts – filled with pits and channels. Far more surface may be available within the mass than on the outside and the interior is often covered with creatures that filter food out of the sea. A five-kilogram piece of marine stone may hold a thousand tiny worms.

The substance of the reefs is also under constant attack by the waves. The rubble rolls into deeper water and must be replaced. As a result, the gothic complexity beloved of film-makers can survive only in relative calm. The finest reefs live in the smoothest seas. The Red Sea has few storms and contains many intricate and beautiful structures. A long fetch of open water in which surf can build up, or a plague of cyclones, makes for a simpler design. In the stormiest places just a few dull mounds survive.

Matthew Flinders saw but a tiny part of the real biological diversity of the Barrier. Now we know far more. Thousands of species of coral have been described: sea whips, sea fans, domes, brains, organ pipes and others. Some will cross with other kinds in an aquarium. The genes show that many – perhaps most – can mate with their close relatives to make new hybrid forms. The main group of stone-builders, once believed to consist of fewer than two hundred species, is a mosaic of unstable hybrids. As a result it is hard to work out how many kinds exist in reality or whether the idea of a species as a distinct genetic entity applies beneath the waves in quite the same way as it does above.

The capital of the underwater empire forms a triangle that stretches from Malaya to the Philippines to New Guinea. The Atlantic is far less diverse but has its own minor metropolis in the southern Caribbean. The East Indies are rich indeed. Three thousand of the 3700 species of reef fish found across the Indian and Pacific Oceans are found there and more than nine-tenths of the corals of the Eastern Pacific have migrated from that centre.

A quarter of all sea fish live on reefs. They range from groupers that weigh in at four hundred kilograms to others a centimetre long. Animals may be present in enormous numbers. A certain atoll in the Tuamotu Archipelago in French Polynesia has five million giant clams, each up to a metre across, in a lagoon little more than a kilometre wide. More than a hundred thousand different species have been recorded in such places, but that is no doubt a great underestimate.

Their home has a close relationship with the surrounding sea. Many reef-dwellers have juvenile forms that float for weeks in the far ocean before they return (if they ever do) and some fish find their way back not at random but with the help of the clamour made by their fellows as they blow gas bubbles from their mouths and anuses and by shrimps as they click their claws.

An atoll is full of life, but around it the seas are sterile. The clear blue waters of the South Pacific are empty. Darwin noted the paradox that the hotbeds of existence on his lagoon islands, some of which sustain two tons of fish per hectare, are in the middle of a watery desert. An atoll is, as a result, a largely independent system that recycles much of its own production. Each day the polyps pump out almost half their efforts in the form of thousands of litres of mucus that drifts in the inner lagoon, trapping huge numbers of algae and bacteria. It settles to the bottom and is re-used by the hungry inhabitants before it has a chance to wash out to sea.

As in many clean-living places, life on a coral cay is hard. Each step in a food chain wastes precious energy. As a result there can be just a few levels between the lowest producer and the top consumer. Most locations on land have four or five different kinds of creature between the plant that soaks up sunshine and the tiger that feasts on flesh. A place as hungry as an atoll can afford no more than three. So stringent are its rules that each piece of fish faeces may pass through five more fish before it reaches the bottom.

Their simpler fellows are also in constant search of a meal. Polyps suck in tiny items of food, sometimes with the help of flat fans that filter the water. The coral animals in turn are eaten by a variety of small invertebrates and by fish who themselves are attacked by their larger brethren. The prey respond to an evolutionary imperative: live fast, die young and squeeze in the best – and cheapest – chance for sex. The power of competition is stark. The smallest known vertebrate is a Barrier Reef fish, as is the backboned animal with the shortest life span, at a mere eight weeks. Off Okinawa lives another fish that breaks the vertebrate record for the time taken to reach sexual maturity.

The undersea battles take place in many ways. Some are obvious, as a fish devours another or a starfish sucks up a polyp. Others are slower but just as bitter. Some corals form crusts that grow over their rivals and kill them. As a colony encroaches on its neighbours, war may break out. Each carries inherited clues of identity on its surface. When they touch battle begins. At once the warriors attack with stings and poisons to keep the competitor at bay.

In the tropical shallows, biology is bad but physics is worse. Shade is a lethal weapon and any lump of stone that blocks its rivals from the sun as it grows will prevail. Cold, too, can be a killer and many reefs have died when the sea-mounts on which they live were borne out of the equatorial seas by continental drift. Tropical polyps are sensitive beasts and do not survive below 18 or above 34 degrees, while most of them do best at blood heat, at 28 degrees or so. None can grow north or south of a line that passes through southern Japan and southern Texas in the north and through Brisbane and South Africa in the south. Those near the limit grow slower than their brethren at the equator and many an atoll that now basks in warm sunlight is doomed to freeze as the mountain upon which it rests moves towards the poles.

For most such animals, cold and gloom mean instant death. At the end of the second millennium came a great surprise, for a whole new dynasty of stonemasons that thrive in just those conditions was discovered, a kilometre beneath the surface of the northern seas.

In 1998, Britain gained its own stern Caledonian version of an atoll. The Darwin Mounds (named after the ship that found them rather than the scientist himself), two hundred kilometres northwest of the mainland, consist of several hundred separate structures. Now, many more such places have been found. New fisheries in what seemed like featureless parts of the ocean were the earliest hint of their existence. The fish had made their home around great masses of rock on the sea floor.

The rock is made of coral. On the Darwin Mounds, each colony is just a metre or so across. They sit on piles of sand, each the size of a substantial office block, built as liquid seeps out of the sea bottom. Each may bear a population of bizarre single-celled animals as big as a dinner plate. Some colonies are thousands of years old, but their builders have also been found on the legs of North Sea oil rigs, where they must have grown in the past two or three decades. Deep-water polyps are widespread even in tropical seas, where they grow in the chill blackness four kilometres down. The largest discovered so far are off the Lofoten Islands, with another vast outbreak in the seas of Alaska. The colonies may be three hundred metres high and several kilometres across. In the sterile plains of the ocean floor, as in the hungry waters of the open Pacific, the reefs act as centres of abundance for a variety of other forms of life. A new kingdom of the reefs has begun to reveal itself.

Cold-water corals are a reminder that reefs have a complex history, parts of which are quite different from that known to

Darwin. His lagoon islands and those who make them are recent investors in the marine business. A lot has happened beneath the waves in the past three thousand million years. Whole generations of rocky barriers, made by quite different builders, have come and gone. Any entity that stays in the same place and sucks calcium salts out of the sea will sooner or later build a shelter and plenty of bacteria, plants and animals have done just that. Whoever makes them and however solid they look, such edifices have always been fragile. They are the first to suffer in the disasters that have hit our planet but, each time, they have been reborn, sometimes with an identity quite different from what went before. Like Rome, most of today's reefs rest upon the ruins of their past and, as in that city, their remains reflect the ebb and flow of history.

When viewed across the abyss of time, reefs come and go at some speed. The ancestors of the Barrier can be traced back for two million years. The wall of rock has waxed and waned since then, and some has been laid down in the past ten thousand years – long after humans reached the continent. Its fortunes changed as the ocean surged and fell back and certain parts now grow at such a rate that elements of the landscape which attract tourists have been born in the brief years since Captain Cook.

At some periods marine rock covered an area ten times that of today while at others the reefs almost vanished. The earliest reefs were bigger, simpler and grew far more slowly than their modern descendants. Their remains make up vast tracts of the globe's surface – and, in a few places, those who made them still pursue their humble task.

Those first bioherms, to use the technical term for any solid edifice built by sea creatures, were made long before the polyps evolved. Their architects were single-celled beings called cyanobacteria, whose descendants are still around today.

Stromatolites, as those primitive edifices are called, were laid down in shallow seas as sand and silt mixed with the slimy remains of the bacterial clumps. Some are just visible to the eye, while others are as big as a house. The finest fossil examples are in Western Australia. There a bed dated to just 3430 million years ago was laid down in a shallow sea, at or just below low tide level. It was the product of the oldest of all cells – a hint that life itself began not, as often suggested, around some hot vent in the deep oceans, but on the first and simplest reefs.

The stromatolite dominion lasted for more than half the history of life. It reached its peak in the Pre-Cambrian, the period before 540 million years ago, and has gone downhill ever since. The structures grew in the tropics, but the continental plates have carried their corpses to all corners of the Earth. A North American deposit from those days glories in the name of the Gunflint formation. Its rocks are full of iron and silica and make a perfect spark for a muzzle-loader. Large tracts of Russia are made of similar stuff and smaller segments are scattered through the west of England.

In time, almost all the stromatolites faded away although, like Counts of the Holy Roman Empire, a few survivors hang on as faded relics of ancient magnificence in places such as Shark Bay in Western Australia, in the Bahamas and, rumour has it, in freshwater lakes in Minnesota, the details kept secret to protect them from collectors. They were succeeded as marine architects by a variety of other creatures such as algae which are able to combine calcium with carbon and oxygen to make solid stone.

The reefs' next phase came with the Cambrian, that explosion of fossil life. The bacterial masons were helped by sponges and algae that could make hard skeletons and by tiny shell-makers that pulled chemicals out of the sea. As the structures grew they picked up silt that hardened into a formless mass. Those ancient edifices, too, have drifted away from their birthplace to Siberia, Antarctica and elsewhere. On the way, many were pressed and transformed in the depths far below.

The upper Devonian – around 360 million years before the present, when amphibians first made their way to the land – was a high point for the undersea universe. The reefs have never surpassed it, perhaps because their efforts in those salad days locked up such masses of calcium carbonate that their successors were deprived of raw material. The underwater temples buzzed with life. Sharks and armour-plated fish abounded. That happy time for bioherms laid the foundation of millions of square kilometres of the modern landscape of Canada, Europe, North Africa and Australia.

As the Devonian wore on, corals themselves began at last to play a major part. The earliest polyps appear towards the end of the Pre-Cambrian, but they were soft and did not lay down rock. Around three hundred million years before the present the edifices started to look rather like those of today, with corals hard at work and rock piled up at impressive speed.

Then came disaster. The great extinction at the end of the Permian period, 251 million years ago, meant that for long ages the reefs went backwards. The few laid down were once more the product of single-celled organisms.

The penultimate act of the underwater opera came to an end sixty-five million years before the present, with the bang that killed the dinosaurs. It did even more damage under the sea. Many bioherms were destroyed, but they rose, Lazarus-like, from the grave. Once again, the polyps flourished. They were helped by calcareous seaweeds and by foraminifera, tiny beings with solid shells. The climate was just right for masonry. Even better, the continents were so arranged that an ocean stretched around the equator, with North and South America separate and Africa far from the Middle East. Sometimes – like St Peter's on the ruins of a Roman temple – the biological cathedrals rose upon their dead selves, for the mass of limestone made by an extinct atoll or barrier is a sympathetic place to set up home. Its corpse is sometimes fretted by rainwater into a fissured rock where marine life can settle when the seas rise once more.

The decline and fall of the Roman Empire was not a simple affair. It took many centuries, with pauses when matters looked up. Gibbon blames it on the decay of the aristocrats into luxury, the loss of the old religion, the movements of barbarians and even on fights between factions of chariot-race supporters. The record of today's reefs reveals that their history, too, is a messy compromise. Darwin's notion of a lost mountain as a foundation for lagoon islands is correct, but that is not the whole story. An atoll faces not just a floor that sinks but a ceiling that shifts upwards in times of high water – and a basement with a disquieting tendency to move to a new address.

As the Earth heated up, the waters of the last ice age poured into the seas. Taken as a whole, the world ocean has risen since the global chill of twenty millennia ago. At the worst moments of the freeze sea level was around 130 metres below its present level. Britain and France were joined, as were the Americas and Asia at the Bering Strait. The sea has risen ever since, with brief lags and spurts as the weather cooled and warmed – but even if it has been more or less static in the last five millennia, in historical context its waters are still low and have nowhere to go but up.

The 1896 Royal Society Expedition to Funafuti drilled deep, but found only coral. The core torn out of the island's heart was taken back to London and stored in the Natural History Museum, where for a century it gathered dust. Now it has been used to reconstruct the past. Changes in the pattern of deposition of carbon and of other elements show that Funafuti has had a chequered history. The atoll is far older than imagined by Darwin and has had periods of rapid growth matched by others of near-collapse.

Its youngest section has grown with enormous speed since the final departure of the last ice age some eight millennia ago. It has put on twenty-five metres of calcium carbonate since then. For the previous half million years the island grew at a slower rate and took four long breaks as it did so. A closer look at the core shows that its surface was exposed to the weather at each of those intervals, for the stone is etched by fresh water. The seas had dropped and the polyps died – but soon the ice retreated again and the waters surged upwards. The drills of the Victorian engineers never reached Funafuti's basement, but today's sonic probes show that the volcanic plug upon which it rests is a thousand metres down. It sinks at around thirty metres per million years – and the builders can easily keep up with that.

Many of Funafuti's fellows have the same history of stability followed by regime change. For some reason, Bikini and Enewetak each grew more slowly than did the Royal Society's chosen site, while the rocky barrier around Espiritu Santo, the largest island in the Vanuatu group, had a nasty five millennia as the oceans rose, for it found it hard to keep up as the land sank below the waves. In those dark days the polyps made little rock and almost perished. Then the masons reached towards the light, caught up and now flourish once again, a few metres beneath the surf.

Faced with falling land or rising sea all corals have reason to worry, for they must keep up or die. Sometimes they manage and a gradual rise in the level of the water above their heads may even stimulate their growth, but history can sometimes move too fast for even the most energetic polyp. A few tens of centuries ago, around Grand Cayman Island in the Caribbean, the ocean jumped by several metres because of a sudden collapse of Antarctic ice. Its masons, happy in the surf, found themselves in the deeps and died at once. Their skeletons are not alone: hundreds of kilometres to the north, in the Gulf of Mexico, other victims of the same disaster tell their story in pinnacles of limestone a hundred metres down. The Big Island of Hawaii, too, bears many memorials to oceanic calamity. Off its shores are five dead reefs, each deeper than the last, formed over the past half-million years. The most recent is just eighteen thousand years old. Each grew as the sea fell at the time of an ice age, but was plunged into darkness as the climate warmed.

Drills and sonic probes show that, in the medium term at least, the fate of many such reefs is ruled more by the movements of their upper limit – the fatal open air – than of their floor. To add to the uncertainty, over vast ages the sea-mount upon which an atoll makes its presence felt may be borne, with the tectonic plate upon which it sits, to a place too cold for the builders to survive.

The grand simplicity of Darwin's ideas has been muddied by inconvenient fact. On a few points he was in error. Reefs that fringe the mainland are, we now know, often found in geologically active places like the Caribbean and do not, as he insisted, need long periods of stability. In addition, the shallow central lagoons found in all atolls may in part arise when fresh water dissolves the rock, rather than always be evidence of a sunken spike of basalt far below.

Even so, science proves that, for the places he knew, Darwin was more or less right. Oceans rise and fall further and faster than he imagined and the reefs – like life and the Earth itself – are older and more complicated than he thought. Darwin would be gratified to learn that his humble creatures have such a notable past and intricate present. Their history adds lustre to what he called 'the magnificent and harmonious picture' that emerges from the study of such places. Their story is that of the planet, for the earliest reefs were made by the first life of all and the demise of their descendants, so often foretold, may mark the end of existence as we know it.

The portrait of the desert islands has been improved by science, but science – or those who use it – has also sullied their image. The nuclear age vindicated many of Charles Darwin's ideas, but presents a less agreeable picture of ourselves. Many of the places visited by the *Beagle* have been despoiled. Papeete, the capital of Tahiti, where Gauguin set up home, is now a concrete mess. The city's Gauguin museum has almost no examples of the artist's work. Instead it has copies, daubs made in homage to the parody of paradise that Papeete has become.

The city assumed its hideous guise in the 1960s, as migrants poured in from all over the Pacific. They were attracted by the jobs generated by the French nuclear weapons programme. The explosions took place on Moruroa, a thousand kilometres to the east. In the local tongue its name means 'Great Secret' and the military went to some lengths to hide what they were up to.

France joined the attack on the tropical seas in 1966, after Algerian independence put a stop to their nuclear tests in the Sahara. Its politicians had considered underground explosions in France itself or on their Antarctic island of Kerguelen, but in the end the Pacific won. As the General in charge said, 'Polynesia is least exposed to hostile campaigns or worldwide emotional reactions, though we should harbour no illusions in this regard, in view of the test method we plan to use (contamination of the worldwide type).'

De Gaulle claimed that his bomb was 'the most peaceful thing we have invented since France came into existence' but when his plans were announced there were substantial riots in Papeete and plenty of complaints elsewhere. Even so, the programme went on until 1996. France fired off 181 explosions, almost a quarter of which were in the atmosphere. The remainder took place a kilometre beneath Moruroa's surface, in its solid foundations. The authorities insisted that there was no reason to worry and it became a ritual for each passing politician to take a dip in the lagoon to prove its safety. In fact, the International Atomic Energy Authority estimates that half a ton of plutonium and other radioactive material is sealed within the island and that plenty has leaked out into the lagoon. Many people exposed to the tests have developed thyroid cancer which they believe to be due to radiation. Their claims against the French government drag on.

Moruroa's rock was tough, but the blasts caused parts of the atoll to sink five metres as its basement turned into rubble. The authorities denied any problem and even threatened *Le Monde* with legal action after the newspaper published a map of the cracks. Even now, a decade since the last event, the land continues to slump. The island's beaches are dotted with sensors linked direct to Paris to give advance notice of any further collapse. Darwin's sinking mountain has been given impetus by human folly and the coral that grows upon its peak may have a job to keep up.

A thousand kilometres away and a century before, Gauguin learned of the death of his daughter, far away in Europe. The news caused the artist real anguish, and he painted what he saw as his most magnificent and harmonious portrait as a memorial to his child before he tried to commit suicide with arsenic. As he said: 'Into it I put all the energy I could before I die . . . I shall never make a better one or another like it.' The canvas, entitled *Where do we come from? What are we? Where are we going?* measures five metres by two. It shows a Tahitian landscape with an idol, a newborn baby, a young girl, a naked youth and an old man waiting for the end. In the background rises the island of Moorea, its native fauna then still extant.

Science tries, in its pedestrian way, to answer the questions posed by Gauguin and – at least for the first two – has had some success. As for the last, scientists have no idea, but the pessimists among them may wish to turn eastwards from the spot where Gauguin wielded his brush, towards the test site on Moruroa, and despair.

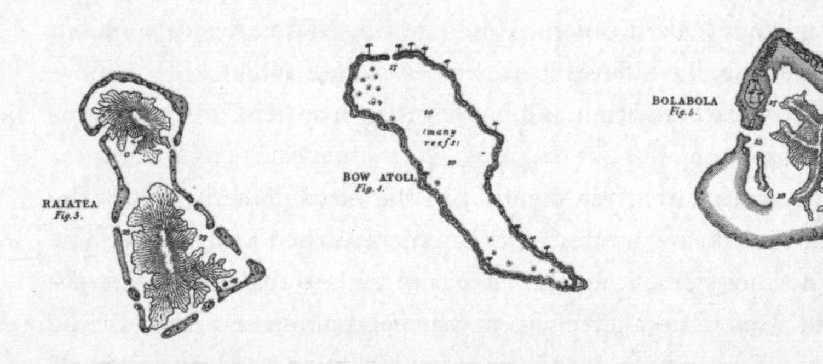

CHAPTER II

THE HYDRA'S HEAD

Perseus, in his courtship of Andromeda, was obliged to decapitate a sea-monster known as the Gorgon Medusa. He placed her grisly head in a bag packed with seaweed, which was at once transformed into coral. For the ancients that legend provided convenient support for their belief that the material came from a plant that hardened into stone when it met the air.

Coral, thought Ovid, was a fine example of a great force of Nature called metamorphosis. His mystical tales the *Metamorphoses* turn on a trick still much used by novelists (in Kafka's most famous sentence: 'As Gregor Samsa awoke one morning from uneasy dreams he found himself transformed in his bed into a gigantic insect'). The process is also of interest to biology as it tries to understand how a baby reinvents itself as an adult and why, in an unwelcome reversal of that process, aged cells may return to the juvenile and unruly form that manifests itself as cancer.

In Ovid's tales death and transfiguration are kin. A live daughter becomes lifeless but precious metal when caressed by her royal father and a narcissistic shepherd turns into a flower as he starves in front of his own reflection. Girl into gold or rustic into daffodil are no more than allegory, but reality is almost as hard to understand. How does a fertilised egg turn into a formless embryo that looks no more complex than a polyp and then into an adult in all its intricacy? How does the body maintain itself against the shocks of everyday life? And why must all of us grow old, depart this existence and break down into a few simple chemicals that go to feed the flowers? Corals give the answer to all those questions, for in their simplicity they retain a talent that Ovid craved. The poet longed for eternal life but, after bitter reflection, accepted that his only hope for such a gift lay in verse. The animals that build reefs have that power in a far more tangible sense.

Like Ovid the Gorgon Medusa lives on in spirit, for her name persists among the Gorgonians (horny corals such as sea fans) and the Medusae or jellyfish. Hercules was obliged, among his Labours, to kill her descendant, the Hydra (who in turn gave her name to a polyp found in fresh water). His task was not easy for the nine-headed beast had the irksome ability to make two heads whenever one was cut off. Not until her central and mortal cranium was excised did the Hydra die.

The Medusa has had her revenge. As Ovid was forced to accept, the gods alone are indestructible and all earthly beings are sure to decline, sooner or later, into the grave. Science now faces the Herculean task of understanding development, from fertilised egg to old age. The twin sciences of embryology and of physical decay – vast as each now is – both began with *Hydra*; and not with the myth but with the animal itself.

That creature and its kin are full of surprises. They reveal some remarkable truths about our own youth and decline. As a result, corals and their relatives are now at centre stage in the struggle to understand some of the basic rules of biology.

For the ancients, in contrast, they were assumed scarcely to be alive at all. The Mediterranean version was thought to be a mere mineral and then, as Pliny put it in his *Natural History*: 'a shrub, coloured green with white berries, which once taken out of water at once hardens and turns red'. It was a lithodendron, a plant that petrified when exposed to air. After a delay of just two thousand years the Italian naturalist Count Luigi Ferdinando Marsigli, who studied the matter in Montpellier, noticed that even underwater its branches were hard, not soft. They bore 'white flowers a line and a half long, with a white calyx from which shot out eight rays of the same length, looking like a clove'.

Marsigli showed the supposed blooms to a young doctor, Jean-André Peysónnel, who had a sudden insight. In 1723 he wrote to the Academy of Sciences in Paris: 'I noticed that what we believe to be the flower of this so-called plant is in reality an insect like a small sea-nettle or anemone . . . I was pleased to see the feet of this nettle move and, having warmed the water where the coral was, all the insects also opened up.' That curious entity was not a marine plant but an animal – and an important part of God's creation, for in those pre-evolutionary days it made a link in the great chain of being that extended from vegetable to Frenchman.

Thomas Henry Huxley described what happened next: 'France being blessed with Academicians whose great function . . . is to prevent such unmannerly fellows as Peyssonel from blurting out unedifying truths, they suppressed him. Peyssonel, who evidently was a person of savage and untameable disposition . . . sent all further communications to the Royal Society of London, and finally took himself off to Guadaloupe and became lost to science altogether.' Voltaire, too, parodied the Academy's attitude to evidence: 'This production called a polyp is less like an animal than a carrot or an asparagus. In vain have we opposed to our eyes all the arguments we have read elsewhere; the classification of the polyp will not change in our times.'

Much as the notion of a multiple beast alarmed the savants of Paris, Peysonnel's theory was soon accepted. Today's comparative anatomy – molecular biology, as it is called – supports his idea. Corals are indeed animals but their DNA shows that they are on a branch of the tree of life quite different from that occupied by men or mice. Sea-nettles are simple indeed, with only sponges – little more than a jumbled mass of cells – lower in the animal hierarchy. The rash assumption that DNA changes at a steady rate dates the split between those creatures and more elaborate beings such as ourselves to a thousand million years ago, long before the Cambrian explosion of novelty when most modern groups are first seen in the geological record. Fossil jellyfish found in southern China, which date from thirty million years before that primal moment, show them already to be a complex and diverse group, with some kinds quite similar to species alive today. Corals and their relatives, the genes and the fossils each prove, emerged long before anything remotely like a human being.

Ancient as their lineage might be, their DNA is not just an incomplete version of our own but contains almost as many genes as that of the so-called higher animals. The common ancestor of *Hydra* and humans, a creature from at least a billion years ago, was rather a sophisticated beast. In some ways it was even in advance of its mammalian descendants, for today's underwater architects retain many genes now found only in fungi and plants and absent from all higher animals, ourselves included. Our earliest evolution, it appears, must have involved gene loss as much as gene gain. The notion that a simple body means simple DNA is, the corals prove, quite wrong.

Fruit flies and nematode worms have long been the workhorses of biology, for they share many genes with humans and can be bred with ease. In fact, the double helix shows that each of those laboratory stalwarts is less like us than was once assumed as – for reasons quite unknown – both have, on the long road from the Pre-Cambrian, lost even more inherited information than we ourselves have. As a result, each is left with a unique and diminished identity of its own and corals have more distinctly human genes than either of them.

The big step forward made by humans or flies compared to polyps is that we have an anus: every one of us, however eminent, is a ten-metre tube through which food flows, for most of the time, in one direction. *Hydra* and its kin are in contrast mere sacs, obliged to suck in sustenance and throw out waste from a single hole. A closer look shows that most animals are, compared to those simple life forms, blessed with an extra body layer. Gorgonians, Medusae and their kin are made from just two sheets of tissue. They are one bag held inside another and stuck together with a sort of biological glue. Higher beings (ourselves and worms included) have a third sheet, sandwiched in the middle. From this grows a set of distinct organs such as muscles, blood, kidneys and more.

In a further advance, flies, worms and men are built around several axes, with a left and a right, a head and a tail and a back and a front. Polyps do have a top and a bottom, but beyond that most of the adult forms appear to have little more than the simple symmetry of a bicycle wheel. The fact that some genes in their embryos are active on just one side of the developing mass may mark a tentative stab at another kind of symmetry, while others that burst into brief activity as development goes look rather like those that determine left and right, back and front and top and bottom in ourselves. Animals with a noticeable left and a right emerge for the first time in the Cambrian era. Their new scheme of life may have been the spark for that evolutionary explosion and the anus followed on. In an unexpected twist to the tale a *Hydra* lives upside down when viewed from human perspective, for the genes that make its lower parts resemble those that build our own heads.

Reef-makers, anemones and jellyfish are united into a group distinct from all others by the presence of unique cells that sting and may kill. Each holds a harpoon with a spring and poison sac attached. The weapon, the most complex structure made by any cell, sits inside a sheath that looks like the inverted finger of a glove. Thousands are found in each square centimetre. A slight touch will fire them off. The poison packs a powerful punch and can injure anybody foolish enough to get too close. Darwin himself, on his brief visit to Cocos–Keeling, was 'a good deal surprised by finding two species of coral . . . possessed of the power of stinging'. For some reason he touched his face with a piece of the underwater stone and found the effect to be as painful as that of a nettle. Some kinds are more potent. Jellyfish kill a swimmer a year in Australian waters and a certain variety no bigger than a fingernail causes the dread Irukandji syndrome, which can cause instant heart failure. Some of the toxins resemble those of snakes, others look rather like the proteins found in certain bacteria that poison food – and some, for no obvious reason, are rather like the surface cues that help human sperm to find their way to the egg.

The harpoon-bearers are known to science as the Cnidaria, a term based on the Greek *cnidos*, or nettle. The adults of the different subgroups are split into six or eight segments separated by internal barriers. Many are colonial and are linked into a sort of shared intestine with mass digestion and a common defence against enemies.

Cnidaria were the earliest creatures to develop a perceptible nervous system, a fact first noted by Louis Agassiz himself. Once seen as a mere random net, their transmission machinery is now known to be arranged into distinct nerve tracts. It even makes a tentative stab at a brain, in the form of a ring of nerves around the mouth. As a reminder of the immense age of the empire of the senses, many of the molecules used to transmit signals through a *Hydra* resemble those that do the same job in our own bodies. Some jellyfish have complex eyes and a clever hunting strategy – and they get tired after a hard day in search of food, for they shut down their nervous system for many hours at a time, to give Dr Peysonnel's favourite beasts the unexpected honour of being the simplest creatures to go to sleep.

Certain kinds have bright colours and, like plants, use pigments to soak up energy from sunshine. To copy themselves, they may bud and spread like weeds, or escape their colony to set up shop elsewhere. Some are asexual, but others enjoy sex, as males and females, or as hermaphrodites. Group copulation is popular. Most of the reef-builders release sperm and egg into the sea over a few night-time hours to increase the chances that the cells will meet.

The moment at which they do so – perhaps just once a year – is set by the length of the day and the strength of the sun. Some species brood their young inside their bodies, while others release them to take their chances in the sea. Often the juvenile forms look quite unlike the adults and live as special larvae that float for miles in the hope of a new home. Many Cnidaria – but not the corals and anemones themselves – include as part of the life cycle mobile forms called Medusae, better known as jellyfish. They may drift for months before they produce larvae that settle on the bottom to make a polyp once more. Anemones and corals reproduce directly, with no intermediate phase.

Whatever their habits, many of those creatures have a talent that others would die for. Robert Louis Stevenson, in his South Seas journal, quotes a sad Tahitian proverb: 'The coral waxes, the palm grows, but man departs.' He was right, for though men leave this Earth in a biological instant, a palm tree may survive for more than a century and a polyp can last for even longer. In Stevenson's day an islander could expect just four decades of life, and the author himself died at forty-four. In stark contrast, some of the labourers whose productions made the beach of Falesa (the title of one of Stevenson's last works) may still survive, a century and more after the novelist's visit.

Age was once a rare complaint but now things have changed. If, in Britain, all deaths before the age of fifty were to be abolished by government decree, the average life expectancy would increase by a mere eighteen months. The aged are so familiar (and with my sixtieth birthday I have joined their ranks) that we forget that many species scarcely bother with them. A turtle given by Captain Cook to the King of Tonga in 1777 lived until 1996. Even certain human tissues in culture long outlive their donors. All living beings face a risk of demise from cold, starvation or disease, but for ageless beasts the danger does not increase as time moves on. For humans in contrast the risk of mortality doubles as each decade passes. A twenty-year-old has a chance of one in a few hundred of expiring within the next twelve months, while for a centenarian the figure is one in two.

Polyps are safe from that grim arithmetic. At the University of Edinburgh some sea anemones collected in 1862 were kept until 1942 when they all perished in an unfortunate accident, but before that unhappy event they showed no sign of a decline in survival or even in fecundity. In principle at least, the *Hydra* I squinted at in school fifty years ago could also long outlive me (in practice, as I recall, they were poured down the sink).

In 1979 the *Guinness Book of Records* interviewed Shigechiyo Izumi, then the world's oldest man, who died seven years later at the alleged age of 120. He and the many other centenarians on the Japanese island of Okinawa ascribed their long lives to the local tap-water, which filters through ancient reefbeds. That claim was much publicised and gave hope to those worried about physical decay. At once, the cynics who populate the alternative medicine industry became rich. Coral calcium, say its advocates, contains the essence of youth and health, and is a universal remedy: 'It's not just cancer. It's all degenerative diseases, lupus, diabetes, MS!' (a statement that was attacked by the United States Food and Drug Administration). The product retails at around a dollar a gram, five thousand times the price of powdered limestone, which is more or less the same stuff. The history of Mr Izumi has given rise to a business with five million customers in Japan alone.

Old age is now a universal affliction. Japan itself has more than twenty-five thousand centenarians. Their lives do not depend on the magic of tap-water, for the people of Okinawa, laden with years as they may be, take in less calcium than do Americans. A proper diet, plenty of exercise and a decent health service all help. Bit by bit the rest of the planet tries to catch up, but even in Okinawa few can hope to emulate Mr Izumi.

The search for the cause of – and perhaps the cure for – the decline from embryo to centenarian was once the preserve of cranks. Now, science has begun to understand why some creatures decay while others do not (although the cranks are still around, with an expensive skin cream called Resilience based on the mashed flesh of Gorgonians). Biologists are hard at work on the thorny problem of why babies are born young and of how two old and tired pieces of flesh can, with a simple gesture, make an energetic new being before they in turn become corpses. Corals help to answer each of those vexatious questions.

Why do all of us deteriorate and die in the end, however good our water supply? The first real clue about immortality – and the key to embryonic development – came when the eighteenth-century Geneva biologist Abraham Trembley cut a *Hydra* crosswise into two. To his surprise both halves grew back into a whole; the body grew a head and the head returned the compliment by sprouting a body. With skilled surgery Trembley could persuade some of his subjects to develop seven heads, almost as many as their mythical namesake. The animal can do much the same of its own accord, for to copy itself it buds off a new individual from the side of its body – which means, in effect, that it lives for ever. Like a potato or a geranium a *Hydra* can split and re-grow more or less without limit.

Inspired by Trembley's work, Charles Bonnet, a fellow inhabitant of the city of Calvin, found that other species could regenerate in the same way. A certain flatworm can rebuild its entire body from a fragment just a hundredth of its original size. Salamanders regenerate tails, limbs, jaws and even eyes, but they are less accomplished than is *Hydra* – a salamander can grow a leg, but a leg cannot grow a salamander. Bonnet suggested, in a remarkable prediction of later discoveries, that such creatures contained 'sleeping embryos', cells that remained ageless until called upon. They woke up whenever a part of the body was removed and replaced the absent piece.

Bonnet's somnolent fetuses are now called stem cells. They are at the centre of a huge scientific effort, for they hold the key to development, to bodily repair and – perhaps – to old age. Their special talent is to divide to produce a daughter with a defined job to do and another that remains forever young, as a stem cell in her own right. Soon, some proclaim, they may summon up a *Hydra*'s talents to change the lives of men and women. Cnidarians provide crucial clues about how such things work. They also hint at why, when they go wrong, they can lead to disaster. The outburst of cellular youth known as cancer is, more and more, seen as a disease of sleeping embryos.

Unlike ourselves, the lowly *Hydra* does not set down a range of specialised organs, but replaces most of its substance at regular intervals. As it grows, newborn material streams through the body, dies and is lost from the head or the foot. Such constant self-sacrifice gives the beast the immortality lost by its namesake when Hercules cut off the sole part of her anatomy that could not be renewed.

Around three-quarters of most cnidarians' bodies are made of stem cells, which can, when called upon, change into nerves, glands and the precursors of sperm and eggs. As a result, such creatures are able to generate a new body from a small fragment and can do so again and again. Their powers of renewal are potent indeed. Box jellyfish – lethal inhabitants of tropical seas – have no obvious brain, but twenty-four eyes. Most are just pits but eight have a lens and retina and look rather like our own organs of vision (which cannot, needless to say, re-grow, or even be repaired). In the box jellyfish, a whole eye can be regenerated from a single cell.

Humans share such talents but not for long. A young human embryo, like an adult polyp, has a unique gift. Every part has the capacity to develop into almost any adult organ. The body has more than two hundred kinds of tissue and, in its first few hours, each section of an embryo can make them all.

We retain that ability until we get to almost the size of a *Hydra*, a creature a few millimetres long. Identical twins are evidence of what the infant form is capable of, given the chance. They arise when a fertilised egg splits. Ten days after sperm has met egg to make a structure with thousands of cells, half a person can – just like half a *Hydra* – make a whole one. Technology shows how versatile the very young can be. Couples at risk of having a child with an inherited disease may use *in vitro* fertilisation to ensure that an unaffected egg is used. To check its status, the geneticist takes a single cell from the eight or so present four days after fertilisation and examines its DNA. If the news is good, the tiny ball of tissue is implanted and, with no apparent difficulty, the diminished object grows into a normal baby.

Embryonic stem cells were isolated from mice in the 1980s and in 1998 were grown from human fertilised eggs. They are now widely used in the laboratory. Most of the raw material is obtained from fertilisation clinics, which use several eggs from each couple in the hope that at least one will survive. The excess is discarded, or joins the thousands held, for no obvious reason, in freezers around the globe. They can also be obtained from the reproductive organs of older fetuses after pregnancy termination, or from young, healthy and altruistic women who give their eggs for the benefit of science.

The crucial material is taken from an inner mass of thirty or so cells found within a sac that burrows into the wall of the uterus. They cannot make a man or woman – they are not totipotent, to use the technical term – for without their protective envelope they can no longer implant into a mother's body. They can however be multiplied in a dish into millions of pluripotent copies, each of which has the potential to make a variety of tissues. Such cells are kept in their pristine state by adding a protein whose normal job is to interfere with the progress of cancer.

As development proceeds, each part of an embryo gains an identity of its own. Infancy ends when a child begins to speak and any child can learn any tongue, from Welsh to Chinese, with about the same degree of ease. Within two years of birth the laws of grammar and vocabulary are fixed in the juvenile mind. Once fluent in a native language it is hard to pick up a second. As the years roll on the more difficult it becomes until, for many of the elderly, language itself disappears into the mists of time.

The passage from egg to adult is much the same. In its first few days, every part of the embryo has the potential to learn any biological dialect, be it that spoken by liver, brain or heart. As it ages, each body element becomes more fixed in its habits and finds it more and more difficult to change the language in which it expresses itself. Embryonic stem cells keep some of their abilities until almost the date of birth, but those of an adult lose most of them. As development goes on, blithe and uncommitted parts of an embryo become stolid and monoglot branches of the body's culture.

Adults do retain a few such cells. They play a crucial part in renewal. Unlike those of a *Hydra*, the sleeping embryos of men and women are rare indeed. A few dozen, tucked away in inaccessible places, are responsible for the care of whole tissues. Each has a series of genes that remain active throughout its life, but are switched off in almost all its fellows. An almost identical set is at work in most parts of a polyp as evidence of their vast powers of regeneration and of the ancient origin of the body's system of repair.

All tissues need maintenance and many organs are restored as a matter of course. The liver is always under construction, as Prometheus noticed when he was punished for the theft of fire by being forced to watch his own grow back each time an eagle ate it. Every day each of us replaces billions of cells in an attempt to resist the ravages of the years. As a result, all of us are younger than we feel. The age of various body parts in a series of middle-aged people was checked using a measure of how much of the radioactive carbon picked up from the tests that blasted the atolls in the 1960s had been lost from each one. The average age of all organs in an adult frame was no more than ten years. No tissue is as sprightly as a *Hydra*, but some turn over at considerable speed. Skin cells last two weeks, the red elements of the blood four months (they travel several thousand kilometres in that time) and even a typical liver has not reached its first birthday. The brain and the heart are more reluctant to renew themselves, and are as a result almost as venerable as the people who bear them.

Most bits of the body can fix local damage. Lost blood is soon replaced, a broken bone is knitted and (with rather less enthusiasm) even severed nerves or damaged hearts can in part be repaired. Such abilities are limited, for we can grow a fingernail but not a finger and can restore bits of a liver but not of an eye.

Replacement and repair depend on the labours of adult stem cells. They retain a limited version of the embryo's ability to differentiate into tissues; those in the bone marrow know only how to make bone, the skin versions turn into skin and so on. Such cells were discovered half a century ago when marrow transplants were first used to save the lives of people with leukaemia. After the transplant the damaged corpuscles were replaced with a new set that carried not the patient's genes, but those of the donor – proof that a small sample of tissue could multiply and restore a failing body part.

Such cells face a relentless challenge, for they must renew themselves until their owner dies. Each is forced to deal with the worst that life can throw at it and has, as a result, evolved to be tough. They possess a series of pumps to get rid of poisons before they can cause damage and are also blessed with a clever mechanism that keeps their precious lineage safe from harm even as their descendants sink into decay. As one adult stem cell divides to make a new version of itself, plus another on the way to becoming a tissue, the lucky daughter destined to retain its special status hangs on to the original DNA strand used as a template when the double helix is copied. Her sister, committed as she is to making a defined structure such as a blood cell, gets the copy. DNA replication is never exact and holding on to the master version allows the stem lineage to cut down the chances of error. Any mistakes are passed to the daughter that becomes a mortal part of the body, while her sister remains relatively pure, reasonably ageless and tolerably able to stick to her job as a factory for flesh.

When life is quiet the special cells may rest, or divide now and again to make the next generation of stem cells. In a burst of activity after injury or other damage, large numbers of new stem cells can be generated before each lineage settles down to make new bodily material. When the crisis is over, many of the lines may lose their unique talents and revert to a mortal state. Like worker bees in a hive, a horde of supporters feed and 'talk' to the cellular queens to ensure they produce the right kind of progeny and stay where they belong. If the helpers are damaged the system breaks down.

The bone marrow makes not just millions of red blood corpuscles every minute, but white cells of various kinds, together with platelets that help the vital fluid to clot. A single cell can reconstitute the entire blood system of a mouse – and the process can be repeated from mouse to mouse, giving a line that outlives any real animal but retains its own almost infinite talents. Impressive as that is, the progenitors of blood have lost most of the joys of youth, for they make only that tissue and not – for example – liver. Why should the polyps be so gifted in comparison and how does a human stem cell, young or old, decide to abandon its status and to advance towards its end?

Once again, *Hydra* gave the first hint of an answer. Its ability to sprout a new head, or a new body, was the key to the modern sciences of embryonic development and adult repair.

In 1905 the American biologist Thomas Hunt Morgan, whose later work on fruit flies proved that genes are borne on chromosomes, became interested in the growth of hydranths, the young polyps of *Hydra*. He suggested that 'a gradient of material is regulating the hydranth . . . from the apical towards the basal end'. As a fragment of body always grew a head and a fragment of head a body there must, he thought, be a system of molecular information based on opposed sets of chemicals whose concentration changed along the animal's length and determined what grew where.

He soon found that an extract of mashed head applied for a short time to another individual caused new heads to grow. Soluble substances were indeed in control. Now we know that both development and repair turn on a series of messages that diffuse from a source to a target. The source may be a master tissue that determines the fate of several slaves, or a local centre that responds to damage. The targets bear receptors upon their surface. As soon as a message arrives the molecule binds to the appropriate site and information is passed through a series of internal go-betweens to the DNA itself. The double helix responds, switches on the appropriate genes and tells the tissue what to do next.

As it grows, every part of every body resounds with news about when and where it should divide, rest or give up the ghost. Each cell is in constant conversation with its neighbours and with its masters. It responds to vast numbers of signals, internal and external, and is festooned with receivers that react to their presence. The signal pathways are arranged – like the telephone, the post and the Internet in a huge city – into separate but complementary systems. They tell their target where it is and where it should go. Commands from another part of the body, sometimes near and sometimes far away, for just a moment or for many years, control an organ's fate. They are carried by proteins, by small nucleic acids or by ions such as calcium (which pleases food-supplement cranks). Often an embryo adopts a new identity at the point at which the concentration of its chemical overlord passes a threshold. A smooth trend in the messenger molecule then causes a sudden change in form, a *Hydra*'s head that sprouts from a sac-like body, or our own neck vertebrae that give way to a quite different construction, the skull.

Sometimes the envoy bears a stark command: to commit suicide. For *Hydra* or human, young or old, in the midst of life there is always death. If you are a fast reader, you have made about twenty billion cells since your eyes strayed to this chapter. All but a few will soon be gone, in an execution known as apoptosis (the Greek for leaf-fall). Cells are slain for many reasons: they are in the wrong place, their DNA has been damaged or their job is done. The change from caterpillar to butterfly or from infant to pensioner calls for planned destruction. Men and women have webbed hands and feet before they are born, but cellular suicide removes those sheets of skin before birth (Cygnus, in Ovid's version of metamorphosis, reversed the process and turned into a swan).

The loss of old body parts, like the origin of new, is under careful control. A jury of genes – whose names, Reaper, Grim and Toll, would be at home in Greek tragedy – decides who will survive and who will not. They pass their verdict down the channels of command and mark their victims with a molecular Black Spot, a label that instructs an enzymatic executioner to cut its target into ribbons.

The killer proteins began among the cnidarians, for they possess such things while their more primitive kin the sponges do not. Each executioner has a segment called the death domain which – as Ovid would be delighted to learn – is identical throughout the empire of existence, so much so that the version that kills unwanted elements of our own bodies is also at work in corals.

Stem cells and their signal mechanisms are a touchstone of modern biology. They have also been promoted as the start of a new era in the healing arts – but an era which, like many such, has not yet quite arrived.

Optimism is the occupational disease of the medical profession. Dozens of illnesses – Parkinson's disease, heart failure, diabetes and more – may, some say, be helped with such technology (it has also been touted as a weapon against wrinkles). The new science might even mark the beginning of the end of the struggle against the burdens of age itself, even if only snake-oil salesmen promise that it will return us to the immortal state of our ancestors.

Medicine often takes a utilitarian view: if it works, use it, and if we have no idea why, who cares? Vaccines, aspirin and psychoanalysis are each effective in their own way although their inventors had no idea how they did the job. For people in desperate need why not just inject the magical material and hope? That approach is simplistic, even brutal – but it worked with heart transplants. Why should it not do the same with simpler subjects?

The lamentable history of gene therapy, in which the pledge was made – but never met – that DNA surgery would become routine, should be a warning. Even so, stem cells are biology while DNA is chemistry; cells have evolved into living entities that multiply of their own accord, while nucleic acids must bow to the stern laws of the physical world. A torrent of hopeful claims about their potential has been matched by an avalanche of cash.

Some stem cells are quite adaptable, given the chance. In the laboratory those from mouse hair follicles can turn into nerve or muscle and will even grow into simple nerve fibres when transplanted back into mice. In rabbits, too, an injection of such material a long way down the road from stem status allows a damaged muscle to repair itself.

They already help in treatment. Chemotherapy and radiation attack parts of the body that divide fast, which is why they are used against tumours. At the same time they play havoc with active natural tissues such as bone marrow. Now, marrow stem cells are injected, while conventional treatment is continued to help put matters right. Whatever its promise, any more sophisticated use of such material is in its earliest stages, even if some people have been fooled, at some expense, into untested and ineffective therapies.

The new form of medicine faces many practical problems. A patient's own cells are the most desirable, but for some people a donor is needed. That raises the problem of rejection. Cnidarians are the simplest life forms to be blessed with molecular cues that enable them to identify foreigners. Our own immune system has a rather similar genetic basis, but is far in advance of theirs.

Its exquisite sensitivity is bad news for those who hope to inject alien material into a failing body – but the young give new hope. In our first days we are more tolerant than we are later in life. Like a babbling baby, a newborn's blood contains a few cells in transition from youth to age. They have not quite mastered their assigned tongue and are more advanced than those in the embryo, but more primitive than bone marrow. As juvenile tissue sparks off only mild immune reactions when transplanted, umbilical cord blood is easier to accept than is that from adults. It was first used to treat a child with inherited anaemia and such transplants are now common.

Another approach is to use patients themselves as a source. Heart failure kills millions of people each year and cripples many more. After the attack, cardiac muscle dies and the organ loses its ability to pump blood. Survivors – and Britain alone has more than a million – may gasp for breath and find it impossible to walk. The heart has few stem cells of its own, but after a cardiac crisis their equivalents in the bone marrow may wake up and move to the site of damage to try to put matters right.

Mice with heart failure given injections of marrow can repair three-quarters of the dead tissue. Some human heart attack survivors, too, show a real improvement after an extract of their own marrow is injected into the organ, or into a vein in the hope that it will move to the target. Quite how this acts is not clear. Such material might reinvent itself with a new identity, or fuse with local cells or make signals that tell the tissue to repair itself. The mechanism is obscure but, the surgeons assure us, the medical effects are real.

Many conditions in which tissues must be restored – Parkinson's disease, multiple sclerosis, recovery from surgery and more – could benefit from this approach. Even severed nerves can, some say, be persuaded to rejoin with the help of stem cells taken from a patient's own nose, where nerves – unlike those in the rest of the body – often renew themselves. The precursors of sperm are also a potential source of such material.

In spite of the optimism the general use of stem cell therapy is still no more than a hope. Sleeping embryos live in inaccessible places, divide but slowly and need complex messages to do their job. They are hard to culture and keen to abandon their flexible state. In addition, each bears within itself a terrible threat.

This emerges from the tie between the fate of organs as they develop and that of their owners as they slide from cradle to grave. Programmed cell death – mortality foretold – is as essential to life as is the birth of new tissues. Degenerative diseases happen when a lethal messenger sends too insistent a signal and a segment of the body falls into an undue desire to end it all. An ill-advised decision to stay alive is even worse, for that leads to the second-biggest killer, after heart disease, in the Western world: cancer.

The sinister power of the embryo shows how real the risk can be. Material from that source appears to be ideal for use in medicine. All that might be needed is to place it near its target, where it should respond to local signals and develop in the appropriate way. However, its enthusiasm may go too far and lead to disaster. An extract of embryo injected into mice often sparks off cancer. Certain human tumours, too, arise when eggs or their precursors – the ultimate stem cells – wander into the wrong place. There they develop into unwelcome structures such as teeth or bone that may grow into tumours and kill. An excess of youth, it seems, is as fatal as an overdose of old age.

The notion of cancer as stem cells gone wrong came from the observation, forty years ago, that each of the billions of guilty elements in certain malignant blood diseases shared the same chromosomal error; proof that they descend from a single progenitor out of control. The idea that one such structure could multiply by millions was tested with an experiment that would be rejected by any ethics committee today. Sections of a tumour were injected back into a patient to see if it would cause cancer at a second site. To succeed (if that is the right word) more than a million cells were needed each time, proof that a tiny fraction of its mass – the stem cells, as we now call them – was responsible.

Such rogue behaviour marks a move back into the evolutionary past, for the malignant entities act like a *Hydra*, dividing not under central control but at a rate limited only by the resources available. As the disease takes hold new products spew out of the tissue factories and refuse to commit suicide on demand.

Healthy cells age and die even when removed from their natural habitat. Those cultured in the laboratory divide for a set number of times and then stop; and the shorter the life of the species from which they come, the sooner they do so. Most of those taken from our own bodies go through about fifty divisions before they give up the ghost. Some, though, act as an unwelcome reminder of the dangers of immortality for they refuse to give up when requested. One such was discovered half a century ago, in a tissue sample from an African-American called Henrietta Lacks. It still thrives in the laboratories of the world, which now hold far more of her bodily material than did Henrietta Lacks herself. She, alas, has gone, for she died of cervical cancer in 1951.

The tissue sample came from her tumour. The cells have returned to a condition in which they divide without limit. For their unfortunate donor – and for millions of others – their hunger for a permanent childhood led to disaster.

Birth and death are busiest in those parts of the body that – like lungs and livers, or the whole of a *Hydra* – call for constant renewal. Each depends on its stem cells to repair the natural shocks that such flesh is heir to and each is more liable to cancer than are more stolid organs, such as the heart, in which tissues are almost never restored. Cancer is repair gone mad; a nightmare in which

Charles Bonnet's sleeping embryos stay awake long after they should have returned to healthy slumber.

Certain drugs can suppress its symptoms for years but, too often, a break in therapy leads to a resurgence. Like the Hydra as she sprouted her mythical heads however often they were removed the tumour recurs, for the essence of the problem, the stem cells, remains at fault. Even as treatment continues they continue to pump out extra copies of themselves, together with vast numbers of unwanted daughters. Because they have evolved to resist the shocks and poisons of everyday life, they can cope with the chemicals used to kill their more fragile descendants. As a result, the crucial elements persist even as therapy goes on.

The fault that tips a cell into a wakeful and cancerous state can be caused by chemicals such as those in tobacco, by radiation or by a virus (which is what did for Henrietta Lacks). It may also happen by simple chance, or be the result of a mutated gene inherited from a parent. Whatever agent is to blame, damage to a piece of DNA that makes or responds to signals is often involved.

Cancer of the colon is the third-commonest form of the illness (after breast or prostate cancer for women and men respectively, and lung cancer). One person in twenty is diagnosed with the condition at some time. The biologist J. B. S. Haldane (who left University College London for India after a row about library hours and who was a major figure in the study of the evolution of ageing) wrote of his own experiences in a memorable poem not long before his death: 'So now I am like two-faced Janus,/The only god who sees his anus.' The poem, 'Cancer is a Funny Thing', begins: 'I wish I had the voice of Homer,/To sing of rectal carcinoma,/Which kills a lot more chaps, in fact,/Than were bumped off when Troy was sacked.' Haldane's new insight into his rear end came from a colostomy, but his bowel cancer had already spread, and soon killed him. Now, science brings new hope.

The inner surface of the intestine, the organ that marks us off from the polyps, is a hectic place, for most of its lining is replaced every five days or so. Its lower section is covered with vast numbers of finger-like protrusions that absorb food and water. Among them are scattered millions of pits, each of which has a minute nucleus of about half a dozen stem cells at its base. Their job is to restore the damage suffered as the body's chemical reactor does its work.

Each pit makes hundreds of new cells a day. They pour from their factory and migrate upwards, dividing as they go. Soon they pass beyond the reach of their master's commands, cease to divide and join the gut's protective sheet. Within a week they are pushed forward to the tip of a protrusion and, in answer to a call from the DNA executioners, they die. The brief life of the gut lining is a microcosm of development itself.

A certain cancer of the lower intestine runs in families. The masters of renewal hidden in each intestinal pit break into revolt, leading at first to the formation of harmless lumps of tissue called, in a macabre echo of the undersea world, polyps. Unless they are removed they can spread and kill.

A single copy of the damaged gene is inherited from a parent. It does no harm as long as the second, normal, version provided by the other parent continues with its job. Should that backup be hit by mutation – which, given the untold millions of divisions in the gut, is bound to happen sooner or later – the machine runs out of control. The pits pour out daughters who refuse to obey the order to commit suicide. In normal circumstances they survive for just a couple of weeks, but their new and extended lives may last for years. More and more genetic errors build up until the diseased tissue becomes malignant.

The fault lies within the signal network in charge of cell division. The system involves dozens of steps that pass on, or block, a particular message. Even so, it has some pivotal points. A mutation called *wingless*, once no more than a laboratory curiosity among fruit flies, is close to the centre of control. Flies unlucky enough to inherit that error stay on the ground, for they lack the physical means to get into the air.

What seemed a discovery of interest only to entomologists has become the key to an intractable illness. The new ability to search out matching lengths of DNA in different species shows that the genes that help build the flight machinery of a fly are also found in ourselves. We have no need for wings, but the undamaged form of the *wingless* gene (or rather genes, for *wingless* is part of a family of related DNA sequences) is at the centre of our machinery of repair. The crucial section of the double helix and its fellows, when they go wrong, can lead to cancer in the intestine and many other places.

The gene makes a messenger, a major player in the network of development. When it works as it should, *wingless* keeps stem cells under control. After a scratch or other damage it puts out a special protein, the first in a chain of signals that tell those talented structures to divide and put matters right. When *wingless* shuts down its protein envoy disappears and division comes to an end. As a result, bodies are cared for and wounds healed as they should be.

The colon's control system is ancient indeed, for our *wingless* genes are almost identical to those that leap into action when *Hydra* buds a copy of its own body. Indeed, the message that promotes the growth of that creature's head is so similar to that of vertebrates that when the *Hydra* version is applied to toad embryos it persuades them to make an extra head of their own. Humans have a dozen members of the *wingless* gene family and cnidarians are the same. Fruit flies, in their reduced state, have but six. In the billions of years since the split between people and polyps some parts of the molecule have changed, but its crucial section, the sequence that keeps stem cells under control, is almost unaltered.

Sometimes a mutation persuades the *wingless* gene to refuse to give up when its job is done. Its DNA control stays in the 'on' position even when it has been told to switch off. Repair goes on long after it is needed and unwanted products continue to stream out of the biological workshops. In time, the symptoms of cancer make their presence felt.

Not all the news from the intestine is bad. Often an operation can save a patient, and for the inherited form of gut cancer that treatment is now routine. Once, a diagnosis needed physical signs of trouble and sometimes they came too late. Now the altered gene can itself be used as a first sign of the problem, well before any physical changes can be seen. Medicines that interfere with its signals, with surface receptors, or with the chain of command that passes orders on to the DNA are already used in treatment. A few are even based on extracts of cnidarians themselves.

Cancer has another tie with a *Hydra*'s ageless state, for it is overwhelmingly an affliction of the old. Their tissues have been restored until they can be restored no more and their repair machinery is stretched to breaking point. As a result, almost everyone over seventy has some form of the disease. The most frequent age of diagnosis for the colon form is seventy-two, and half of that age group has an intestinal polyp. The condition is scores of times more common among the elderly than among teenagers. As the stem cells fail, rejuvenation gives way to the ailments of advancing years.

Medicine can stave off some of the symptoms of age, but not its inevitable end-point. We are ahead of the coral-builders in many ways, but those simple beings retain a talent that we lost long ago, for they have, at least in mitigated form, the gift of eternal youth.

Ovid would no doubt find a moral in our fate, for it points to an existential dilemma. Life has made a Faustian chemical bargain in which the excesses of youth are repaid in the currency of old age. The guilty party is oxygen. We cannot live without the gas for more than a few minutes but cannot live with it for more than a few decades.

The element made its first appearance in gaseous form as a by-product of the manufacture of stone by the earliest reefs. As a result, *Hydra* and its predecessors in their endless prime impinge upon our own lives, for human decay can be traced back to their own discovery of a new and almost immortal way of existence.

Three billion years ago, oxygen was a vicious poison. Almost the whole of the global store was locked into a firm alliance with other elements, carbon included. The vital element began to enter the atmosphere as the builders of the ancient stromatolite reefs combined the sea's calcium salts with carbon dioxide from the skies. At once, the gas was liberated from its bonds.

Their triumph is set in rust – iron combined with oxygen. In its earliest deposits the metal is found combined with sulphur to make a yellow and glistening material called pyrite – 'fool's gold' – or is associated with a single atom of oxygen in a greenish mineral. Then came a change. Later ores are red, because the iron picked up the oxygen atoms that had by then reached the atmosphere. A subtle change in the chemistry of the black shales of South Africa shows that our planet took its first gasps just 2322 million years ago. After the Great Oxidation Event there were some swings in the abundance of the gas, with a peak of 30 per cent – a third more than today – around 275 million years before the present.

Oxygen seems benign enough, but it has a darker side. Now and again it falls into an unstable form in which a free electron armed with an electrical charge buzzes around its atomic core. Large amounts of that hazardous material leak out of the molecular factories that generate energy from food. Reactive oxygen species, as the noxious products are called, damage DNA and proteins and attack membranes. A typical cell gets ten thousand hits a day and can survive only because a series of repair mechanisms has evolved to patch up the damage. We take several breaths a minute to fuel our lives, but a hefty proportion of our energies goes to fixing the ravages of the gas that kills us even as it keeps us going.

Reactive oxygen has a strong tie to cancer, for carcinogenic chemicals such as tobacco, ultraviolet light and X-rays all generate lots of the stuff. Even those who avoid that illness will suffer from oxygen's malign effects sooner or later, for the element in its degenerate form causes most of the afflictions of the aged. Deafness, diabetes, Alzheimer's disease, feebleness, obesity and many other pains of the years are all side-effects of that ambiguous material. Diverse as they may appear, the conditions that kill us off all share the same chemistry.

Evolution has come up with many mechanisms to reduce the damage, with dozens of enzymes that break down the poison and cut out, repair or replace battered pieces of DNA. They do their job quite well (and those rare children born with missing bits of that machinery die very young) but sooner or later they reach their limit and we pay the price.

Power stations suffer the worst damage in their furnaces, where air combines with coal or oil. We burn most of our own fuel in cellular structures called mitochondria. Like industrial boilers they depend on cheap fuel and plenty of oxygen and, like them, they wear out quickly.

The first steps in the body's energy-production machinery evolved long before mitochondria appeared on the scene, in the days when free oxygen was rare. They still do not need it. The later stages, in contrast, depend upon the vital gas. As the protein wheels of the mitochondrial factories grind and spin, they pump out vast quantities of reactive waste.

The mitochondria pay a heavy price for their inefficiency. Their machinery becomes more and more battered with the years, and they become bloated, obese and inactive. The problem is at its worst in the body's least *Hydra*-like tissues, the brain and the heart, where cells almost never divide and where, as a result, mitochondria are seldom renewed. Slowly, toxic by-products build up. People with chronic heart conditions, or Parkinson's disease, may have two thousand times as many damaged mitochondria in their failing tissues than is normal. As further proof of the role of such damage in physical decline, an increase in the mitochondrial mutation rate in mice leads to premature old age.

As the mitochondrion takes hit after hit it sends out signals that persuade its host to enter the programme of planned death which, when controlled as it should be, is central to life itself. Soon, cell after cell commits suicide. As they immolate themselves the body begins to abandon hope: the testes dry up, the bones get thinner, the brain loses nerve cells, the heart falters until, at last, a bent and weakened frame totters towards the grave. Oxygen addiction – a craving that began before the Cnidaria evolved – has become a lethal habit.

The life and death of mitochondria is not simple. They have some rather ineffective repair machinery of their own and are also helped by the many nuclear genes that protect them from malignant oxygen. Klotho was one of the three Greek Fates who spun life's thread (later to be cut by her grim sister Atropos). Her name has been hijacked for one of those protective genes. It persuades mitochondria to break down reactive oxygen before it has a chance to harm them. In mice, damage to *Klotho* leads to a whole series of ailments similar to those found in aged humans – and an engineered and efficient version of the gene, a Fate who works harder than normal, increases the animals' life expectancy. In humans, the gene comes in two variant forms, one powerful, the other weak. The one in thirty of us unlucky enough to bear two copies of the feebler version suffer more from illnesses such as heart disease and die younger than average.

Whatever the Fates do to help, human mitochondria are not very good at patching up oxygen damage. As a result, their genes soon accumulate errors. Some are harmless and are passed down the generations. Mitochondria are transmitted only through women. The mutations build up at some speed over the years, and the miles, to relive the history of the female line. Two decades ago, the celebrated 'Mitochondrial Eve' became the introduction to a whole science of female geography.

Not, though, in corals. Attempts to trace a cnidarian Eve have failed, for many kinds of polyp vary not at all in mitochondrial

DNA, from individual to individual, or from place to place across wide oceans. Their energy machinery is, so it appears, highly resistant to change.

Coral's genetic conservatism is a result of the efficiency of its DNA repair enzymes. The reef-builders face an even worse problem with waste gas than we do, because as their solar-powered factories soak up energy from light to make limestone they spew out vast amounts of reactive oxygen. To cope, the polyps have evolved protective systems far more effective than our own. At dawn an anti-oxygen shield is put in place within each mitochondrion, ready for an assault by the formidable by-product. It is sturdy indeed. Among the stone-makers, mitochondrial mutations build up at just a hundredth the rate found in humans. Their ability to purge such errors explains both why their female lineages look so similar and, perhaps, how they manage to stay forever young.

To have any hope of reaching that agreeable state we too need a defence against reactive oxygen, and medicine has set out to discover it. Certain drugs interfere with the stuff, and statins, those magical remedies against heart disease, slow down the rate at which it is made. The search is also on for new drugs that might combat its noxious effects elsewhere in the body.

Drugs may help to repair oxygen damage, but the hope of the real visionaries of immortality is to defeat the lethal gas that makes us old even as it keeps us alive. The search for the cure for age goes back to long before the element itself was discovered and many, various and usually spurious are the elixirs of youth that have emerged. The vendors of coral calcium have been quick to jump on the band-wagon and about a third of all Americans now take an antioxidant supplement. Most such nostrums are useless, but a few do have some effect. Although once they seemed quite unrelated in how they gain their power, we now know that most of them act to cut down the damage caused by oxygen.

There is more to long life than what you eat, for the diet of

Hanna Barysevich (who died in 2007 at 118 as one of several claimants to be the planet's oldest citizen) was based on gherkins, vodka, pork fat and potatoes. Even so, food can make a real difference. To stay vegetarian for twenty years adds about four years to life. Tomatoes contain potent antioxidants, as do carrots, fish oils, green tea, olive oil and red wine. Exaggerated claims have been made for such a diet but nobody denies the tie between high-fat, high-protein – and high-reactive-oxygen – food and illnesses such as cancer and heart disease.

One way to cut down the damage is to eat less. Narcissus as he starved by his pool rather overdid the job, but self-denial, within reason, is good for you. A starvation diet reduces the amount of chemical fuel that courses through the body. Its factories pump out fewer pollutants and smash up less of the local DNA. Hunger cleans up the cell just as a slump cleans up the skies. In rats and mice it increases life expectancy by a third and more.

Its effects in humans are more ambiguous. The Japanese, famous for their life expectancy, eat about a fifth less than the people of most other developed nations, and the inhabitants of Okinawa are even more restrained. Experiments to establish whether mild starvation might increase our ability to survive will take many years to complete, but evidence of the joys of abstinence has now begun to emerge. A group of Americans of normal weight volunteered to cut their food intake by a quarter. They lost a few pounds and within six months each had a lower body temperature and less insulin than before – both an attribute of the long-lived. Even their DNA showed less damage than average, perhaps because less reactive oxygen was made.

Other defences against that toxic stuff can help. A smouldering fire pumps out more poison than does a blaze and a mitochondrion with its waste-pipe blocked makes more by-products than one in good condition. Exercise increases the efficiency with which the cell burns its fuel, reduces the amount of reactive oxygen and helps people to live longer.

Cold tunes up man's machinery. *Homo sapiens* evolved in the tropics and moved to chillier climes around a hundred thousand years ago. As our ancestors made the journey, evolution worked to improve their central heating. Ice and snow make large demands for energy and the mitochondrial generators must provide it. As a result, mutations that allow them to work better in the cold give their bearers an advantage. They soon spread. Many people in the colder parts of the world bear mitochondrial types that are rare near the equator. They generate more heat (and make less poison) than do their tropical ancestors. That keeps their owners warm and might also protect against the malign effects of age. A certain cold-adapted form is common in Japan, parts of which are icy in winter. The variant is more frequent among Japanese who survive into their eighties than in the nation as a whole.

The first scientific evidence of a tie between reactive oxygen and corporeal decay goes back to a time before the compound itself was identified. It was stumbled upon in the 1960s by a colleague of mine at University College London. Alex Comfort was the author of a classic text on ageing. In an early hint of the dangers of the vital gas, he showed that artificial antioxidants extended the lives of mice (although he had no idea why). On the sole occasion we met we talked of the shell pigment of dog-whelks, used by the ancients as a dye. He had worked on its biochemistry, I failed to find it in land snails and I never saw him again.

Comfort went on to write another technical work entitled *The Joy of Sex* (the only time I have seen it on household display was on the bedside table of a senior politician with – to my considerable alarm at the time – several bookmarks stuck between the pages). His shift from Tyrian purple to purple passages paid massive dividends, with eight million copies sold, and soon he went off to California to practise what he preached.

Like Ovid, Alex Comfort saw that Eros and Thanatos are kin. Death, he realised, is a sexually transmitted disease. Men die younger than women for many reasons, but a shortage of that powerful antioxidant known as oestrogen is among them.

Cnidarians give vivid proof of the importance of sex to life's end. Many species exist in two forms, an immobile polyp, which is more or less immortal and reproduces by budding, and an itinerant jellyfish, or Medusa, which bears sex cells. The lubricious Medusae are more complex than their lumpish and asexual progenitors. Some weigh a ton and have muscles, nerves, eyes and tentacles with which they hunt their prey.

The Medusa pays a high price for its erotic experiences. It regenerates far less well than its virginal parent and does not bud. Once sperm and egg have been released the animal ages and gives up the ghost. *Hydra*, too, is immortal only as long as it holds to its innocent way of life and does not indulge in sex. A sudden chill at the end of the summer spurs the animal into sexual reproduction, but as soon as it succumbs to temptation it pays the price and departs this world. Its fertilised eggs then survive the winter to produce a new generation of asexual polyps in the subsequent year.

Many simple forms of life shift back and forth between sexual and asexual lifestyles as conditions change. Stress is often the cue to take up sex. Many of *Hydra*'s fellows in fresh water, from water-fleas to fish, share its seasonal pattern. Parasites, too, often go in for the pastime when attacked by the immune system or by drugs, as do plants when faced with heat or drought. The tie between sex, stress and oxygen poisoning is at its clearest in a certain small green alga. There, the genes that make the decision to go in for that form of reproduction look rather like those that fight off reactive oxygen. Sex itself might even have begun as a defence against chemical damage, because uniquely powerful DNA repair enzymes are brought to bear when sperm and egg are made.

Sex cells and stem cells have a lot in common. Egg and sperm stay forever young, even as those who bear them grow old. Sex separates reproduction from repair. The stresses of life are resolved in a privileged dynasty that makes reproductive tissues while the others keep its vehicle, the body, in good order.

The unholy trinity of sex, age and death was noticed by theologians long before science began. Gods, in general, gain immortality through celibacy rather than starvation. Jesus was born of a virgin and so was the Buddha. The Greek deity Adonis, who saw the light of day in the same cave as Jesus, but long before, also came from a single-parent family. In many faiths, eternal life is also a reward for the renunciation of the sins of the flesh – lust, gluttony and other generators of reactive oxygen included. Many saints have died in the odour of sanctity, a scent said to reflect the sweetness of the soul as it departs the body. In fact, a lengthy fast causes the body to break down its fat reserves and produces acetone, with its syrupy smell, plus a near-lethal dose of oxygen in its reactive form. Like Narcissus, the pious have exceeded the recommended dose of asceticism.

Charles Bonnet was much exercised by the doctrinal implications of his 'sleeping embryos'. When, he asked, a *Hydra* makes an asexual copy of itself, do both bits have a soul and, if so, where does the second one come from? The question led him to argue that cells had an inner essence, which awoke when called upon and was passed to the next generation. He became obsessed with the religious implications of his work and his career ran, like his reputation, into the philosophical sand.

Many countries still prohibit the use of embryonic stem cells even in research because of their origin in fertilised eggs, each of which, in the tradition of Charles Bonnet, is assumed to have a soul of its own. The notion that they might be allowed to grow into adulthood in the laboratory is everywhere denounced. The problem particularly plagues the conscience of President George W. Bush, who has signed well over a hundred death warrants. His ban on the use of federal funds for such research is part of his 'culture of life'; the work 'crosses a moral boundary that our decent society needs to respect'. Even for laboratory research American policy is an ethical mess; Germans accept embryonic stem cells as long as they are not German; while the Vatican has threatened to excommunicate church members who work on such things. Britain is less particular than most countries, but faces pressure from Catholic Europe.

Doctors, of medicine or of divinity, may differ in their views about the wrath of God but are united in the belief that stem cells will transform medicine. Scientists are less certain about that claim but are concerned by the intrusion of theology into the laboratory. Why should their research be inhibited by a doctrinal quibble that began in the days of Calvin – and why does theology see both sex and death as a result of spiritual degeneracy rather than an inevitable consequence of evolution?

The tie between sex and mortality goes back to long before the sage of Geneva. The earliest Greek myths are those of Hesiod, the father of legend. Hydra's horrid ancestor, who transformed her victims with a glance that turned them to stone, makes her first appearance in his tales. The Gorgon Medusa paid a heavy price for sex, for as a punishment for succumbing to Poseidon, the Ruler of the Seas, the gods gave her a single vulnerable head and deprived her of immortality.

Hesiod speaks of a Golden Age when age, death and the life erotic were unknown. Then came an Age of Silver when man gave in to the desires of the flesh and, for the first time, felt the cold breath of senility. It was followed by a time of Bronze, the Age of War, and by an even lower point, the Age of Iron, which arrived in around 700 BC, the poet's own day. That period was marked by lust and degeneration. From his own distant era Hesiod looked forward to an Age of Despair, a thousand and more years ahead, in which infants would be born old, indulge in geriatric and obscene copulation and die before their allotted span.

Even if stem cell research proves a chimera, we mortals who live in the grim, sexual and oxygen-defiled times foretold by the ancients can still hope for some kind of Golden Age beneath the waves, for Eternal Reefs Incorporated of Decatur, Georgia, will, for a modest fee, arrange for our remains to be petrified: for human ashes to be cast into concrete balls and used as the foundations for new coral reefs. For a few dollars more they will add an electronic tag that tells passing swimmers just who is in each ball. The Gorgon Medusa would, no doubt, approve.

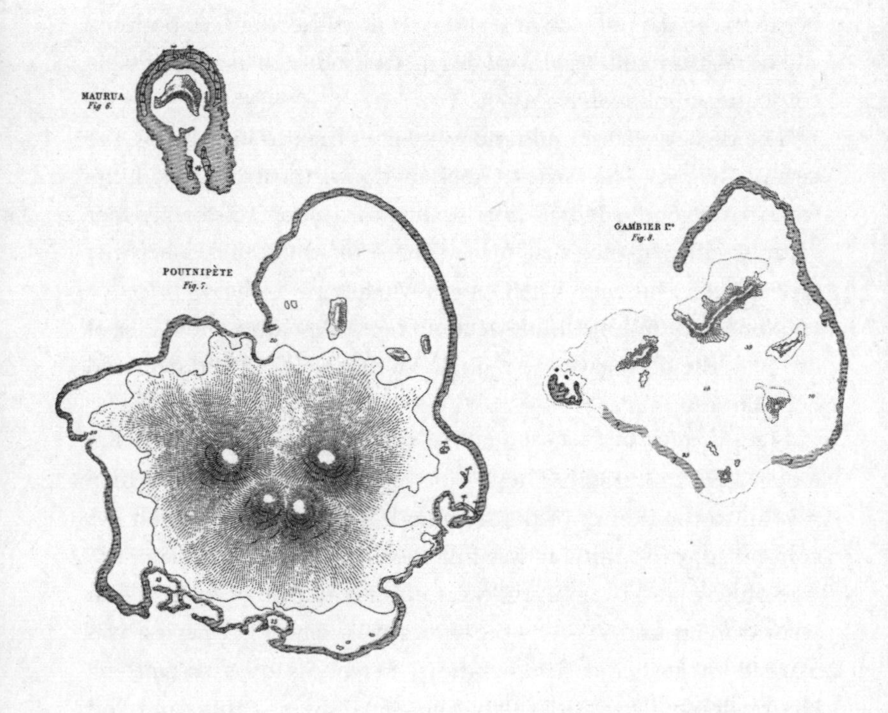

CHAPTER III

THE PLOVER AND THE CROCODILE

In 1871, amidst the rocky remnants of an ancient coral sea, a group of Swiss watchmakers founded a political club. Their ideas came from the philosopher Pierre-Joseph Proudhon (best known for his aphorism that 'property is theft') who had worked as a cowherd in their native Jura mountains. The Jurassic Federation, as the society was called, left a mark on history when it signed up a young Russian Prince, Pyotr Alekseyevich Kropotkin. He saw in its anarchist values a scientific model for civilisation.

Kropotkin was born into affluence but became a radical. After an apprenticeship in the Court of Pages in St Petersburg and a job as *page de chambre* to the Tsar, he commanded a regiment of Cossacks on expeditions to Manchuria as part of his nation's endeavours to expand to the east. On his return he was arrested for his revolutionary activities among the peasantry of the steppes. He fled to Switzerland, but was expelled after the assassination of Tsar Alexander II. The Prince was imprisoned in Russia and, later, in Paris. In time he moved to England and spent thirty placid years in the London suburbs, at 6 Crescent Road, Bromley, not far from Charles Darwin's residence at Downe.

On his arduous journeys and in his prison cells Kropotkin read vast quantities of political theory which, in the fashion of the time, he mixed with biology to form a philosophy of life.

It turned on anarchism, a system of cooperation, of individual rights and – in principle at least – of the decline of all states. In the Jurassic Federation the young Prince found a political home.

In his 1902 book *Mutual Aid, A Factor in Evolution*, Kropotkin put his creed on a scientific basis. He was a passionate evolutionist, but disagreed with Darwin on a central point: Malthus was not an important thinker but a 'malicious mediocrity'. The Prince accepted the struggle for existence but – from what he had seen in his own almost empty land – was unconvinced by *The Origin*'s claim (based, no doubt, on its author's experiences in the tropics) that populations always outgrow resources. Biological conflict existed but, Kropotkin was sure, took place against inclement Nature rather than against other creatures. He had 'failed to find – although I was eagerly looking for it – the bitter struggle for the means of existence among animals . . . On a migration of fallow deer which I witnessed on the Amur and during which scores of thousands of these intelligent animals came together from an immense territory, flying before the coming deep snow . . . I saw Mutual Aid and Mutual Support carried on to an extent which made me suspect in it a feature of the greatest importance for the maintenance of life.'

Kropotkin was not the first to see a benevolent message in the natural world. Two thousand years earlier Herodotus had praised the Nile plover that plucked leeches from between the teeth of crocodiles, to their joint benefit. The great Greek recommended that men should arrange their affairs to follow its example.

His Russian successor went further. He used evolution as a blueprint for society. The more mutually dependent a species was, the higher it stood in life's hierarchy. *Homo sapiens*, at the summit, had evolved to be the most communal of all. Our lust for wealth and for war marked a step down the biological ladder; it was an aberration brought on by a sick and unjust way of life. Inequality and the governments that impose it break the evolutionary rules because they depend on force. As a result, the ideal political system could be found only in places untouched by civilisation.

Captain Cook saw a hint of such an Arcadia on his *Discovery* voyage. The Australian Aboriginals, he wrote, 'may appear to some to be the most wretched people upon Earth, but in reality they are far more happier than we Europeans . . . they live in a Tranquillity which is not disturb'd by the Inequality of Condition.' For Kropotkin, too, the Aboriginals had it right. To go back to their way of life was to obey the laws of biology: 'In the ethical progress of man, mutual support, not mutual struggle, has had the leading part. In its wide extension, even at the present time, we also see the best guarantee of a still loftier evolution of our race.' Once the tyranny of banks and of nations was abolished, man would return to the generous realm in which he belonged.

Others took a different message from the natural world. Thomas Henry Huxley, 'Darwin's Bulldog', in his essay on 'The Struggle for Existence in Human Society', referred to animal life as a 'gladiator's show', a claim described by Kropotkin as 'atrocious'. Friedrich Nietzsche, he of the thick moustache and pitiless doctrine, recommended violence as the natural way forward. He saw in Darwin's work the triumph of the strong over the weak (and one of his own volumes is subtitled *How to Philosophize with a Hammer*). For Nietzsche and his fellow Social Darwinists the rich deserved their good fortune. What use was ethics in the face of the struggle for existence? The feeble might wish to dabble in what Nietzsche called 'slave morality', but evolution's machine had produced a race of supermen to whom selfishness was simple duty.

Nietzsche and Kropotkin were just two of many political thinkers who have picked anecdotes from Nature to support their creed. Even Karl Marx found evidence beneath the waves: 'We see mighty coral reefs rising from the depths of the ocean into islands and firm land, yet each individual depositor is puny, weak and contemptible.' Every atoll proved that collective action, by polyps or by people, was a natural law. Society had been ruined by an altogether artificial medium called cash, which matured into capital and led to exploitation. In an ideal world all would give what they could and get what they needed. In time the state – and money itself – would lose its *raison d'être* and a grand global system of mutual aid would begin.

Kropotkin was, after some years in Bromley, offered a chair at Cambridge, but declined as the university wished him to moderate his calls for universal anarchy. With the apparent triumph of his ideas in the Bolshevik Revolution his Utopia was, it seemed, realised and the Prince returned to Moscow. Within two years he was disappointed and within three dead. Less than a century later so was the Soviet Union, its egalitarian past supplanted by a notably selfish present.

The split between the anarchists and the capitalists reflected a fundamental clash of beliefs. Is humankind ruled by self-interest, or is altruism our true state? What is the lesson from Nature: mutual aid or inevitable strife?

Australia's Aboriginals may have favoured the Russian prince, but Captain Cook also came across plenty of Nietzschians on his travels. The myth of the noble savage had emerged from the accounts of the first French explorers of the Pacific, but in reality the South Seas were full of discord. At the time of Cook's first expedition, Tahiti and its neighbours were in turmoil as the cult of 'Oro spread from its temple on the island of Ra'eatea. There, bodies were used as rollers under the keels of sacred canoes and Captain Cook himself visited a coral pyramid where human bones crunched beneath his feet. Again and again he came across cannibals, although he was generous enough to note that their habit appeared 'to come from custom and not from a Savage disposission'.

Savagery, inborn or through custom, began soon after people arrived in the Pacific. Two millennia ago it took just a few centuries for the first migrants to Mangaia in the Cook Islands to consume the local wildlife. In time they were reduced to eating rats (as the islanders still say, 'It's as sweet as a rat'). Then they began to devour each other. On Mangaia's six square miles lived half a dozen tribes, who fought, fought and fought again, with evidence of forty-two separate wars over fifteen hundred years. Rousseau would not have approved.

Explorers and philosophers may take different lessons from the world around them, but biologists, too, can be less than clear about what is natural. The arguments about cooperation and conflict that split the sages of the nineteenth century have a mirror in modern science.

Evolution – Darwin's 'doctrine of Malthus applied with manifold force to the whole animal and vegetable kingdoms' – is cruel, but can give the impression of being kind. Many creatures do appear to indulge in mutual aid. Such outward generosity reaches a peak in symbiosis, the interaction of different species to what appears to be their joint benefit. Its arrangements are often far more intimate than those of the plover and the crocodile. Reefs are hotbeds of the habit. Certain crabs clean the surface of sea cucumbers and in return find sanctuary in the animal's anus. The boxer crab, in contrast, holds a stinging anemone on each claw to fight off rivals and rewards its mercenaries with scraps of food.

The coral animal's place in such an arrangement is among the best understood of all. The story of how that marine mutualism began – and how it may be close to its end – has lessons both for biology and for those who try to use the living world as a template for civilisation.

Associations between different species, on land or in the sea, were once seen as no more than a curiosity, but now they are at the centre of our view of life. Symbiosis began early in evolution and soon became universal. It seems a perfect instance of Kropotkin's scheme, for each partner appears to gain from the other's efforts and to pay a fair price for its help. The habit was at first referred to – in homage to the Prince – as 'mutualism'. Sadly for anarchists, the more we learn the harder it becomes to separate mutual aid from dogged antipathy. In life, what looks like peace is often no more than an interlude in war and a lengthy liaison may persist only because struggle has run into stalemate.

The semblance of cooperation is all around. Fish bear single-celled helpers that glow on demand, trees depend on the fungi that cluster around their roots and beans flourish with the assistance of special bacteria that draw nitrogen out of the air. Even *Homo sapiens* takes part in the great cooperative feast, for each of us bears within our guts ten times as many bacteria as we have cells of our own. As those on antibiotics soon find out, without them we become constipated at best and vitamin-deficient at worst.

To scientists neither symbiosis nor the struggle for existence has much of a message for human affairs but biology often falls into the lax use of terms familiar to students of society. Ecologists often draw on economics for their metaphors. Malthus' claim that populations are limited by resources gave Charles Darwin what he called 'a theory to work by', but it originated in the eighteenth-century political belief that vice and misery were inevitable, with the best solution to be found in the workhouse.

The laws of the market also help to explain systems in which the proponents appear not to compete in some existential struggle, but to strive towards the same shared end. Sometimes, the market turns to Nature for advice. The first 'industrial symbiosis' was in Denmark, where a refinery, a power station, a gypsum-board factory and a chemical plant were built together. Their by-products, which were once only waste, became raw material: refinery gas went to the power station and sulphur to the chemical plant, while the power station made steam for other industries and ash for the board-builders. On the muddy banks of the Mersey, too, the members of the North West Chemical Initiative – companies who once saw each other as rivals – now find advantage in co-operation as they trade effluents among themselves.

Capitalism, in Scandinavia or on Merseyside, depends on

supply and demand; on the investments made and the dividends received by each partner. Symbiosis follows the same rules. The study of such bargains in biology began with lichens, modest plants that make fine crusts on stones or trees. Their moment came in 1868 when it was shown that each is not one creature, but two. A lichen is an association between a fungus and a single-celled organism then believed to be a kind of green alga but now known to be a simple beast called a cyanobacterium (which is only distantly related to the familiar seaweeds or the green slime often found on the trunks of trees).

At first the link seemed simple. The fungus was in control, with its partner pressed into service. Thirty years later a Miss Potter presented a paper on the subject to the Linnean Society of London – via her uncle as she was not, as a woman, allowed to address the Society face-to-face. She offered a radical new theory of lichens: that the association was not that of master and servant, but of mutual aid based on a reciprocal trade in chemicals. Nobody took much notice and, discouraged, Beatrix Potter gave up science and took up a new profession as the author of children's books.

The benign hypothesis promoted by the creator of Mrs Tiggy-Winkle became the foundation of a novel view of life. The habits of lichens and their many fellows are not, as she assumed, a curiosity, but a major force in evolution. A fifth of all fungi live as lichens and many polyps indulge in a similar arrangement. At a deeper level, the pastime is universal.

It began long ago. Off the coast of Santa Barbara, that bland home of California's super-rich, lies an underwater basin six hundred metres deep, protected by a rocky sill that allows little movement in or out. Its waters resemble those of the Earth in its earliest days, with almost no oxygen and plenty of sulphur and other minerals.

That vast and stagnant bath-tub is awash with mutual aid. The mud is filled with bacteria able to use sulphur as food. Upon them feeds a variety of unicellular beasts and small worms. Almost all have come to an agreement with the bacteria, which lie in sheets beneath their skin and even enter cells. There, they act as chemical factories and in return find a place to live. In the earliest oceans such joint ventures must have been common. In a 600-million-year-old Chinese marine bed lived fungi with foreign cells within. The fossils look rather like modern lichens and their undersea equivalents. Symbiosis, they show, began in the sea.

For a time, all such liaisons looked as amicable as those of Santa Barbara or the large intestine, but as we learn more the shadow of Nietzsche looms larger. Some alliances have lasted for millions of years, but each member of the team still puts its own interests first whenever it gets the chance.

Polyps are a perfect illustration of how precarious such marriages can be. Their relationship with another party is ancient indeed, but even after a vast period of apparent amity it can end in divorce, as conditions switch to favour one partner rather than the other.

The chemistry of atolls has a lot in common with that of the industries of Merseyside. Darwin remarked on the efficiency with which the coral animals extracted salts from the sea to make their raw material. They do the job in much the same way as we make bone: 'The organic forces separate the atoms of carbonate of lime, one by one, from the foaming breakers . . . Thus do we see the soft and gelatinous body of polypus, through the agency of the vital laws conquering the great mechanical power of the waves of an ocean, which neither the art of man nor the inanimate works of nature could resist.'

Most of the work takes place on the animal's surface, in special cells that pump ions of calcium, with their positive electrical charge, from the sea into the cell. Radioactive calcium added to the water enters the polyp's body within a couple of minutes, which is ten times faster than our own kidneys can move the same stuff. Not just dead rock is made, for without the minute fraction of protein present in each block the work of construction stops.

Just like bone, new coral contains many minute pores into which its maker's cells penetrate. So similar is the material to our own skeletons that it can be used to make grafts. Our cells use the sculpted insert as a scaffold upon which to build new tissue, in a surgical symbiosis between man and marine animal.

Now we know how the coral animal's soft and gelatinous self does the job. Another party, unknown to the young naturalist who so admired their talents, has a central role in their battle against the breakers.

The green *Hydra* was once supposed to belong to a new kingdom of life: animals that made chlorophyll, until then assumed to be synthesised by plants alone. The Scots polymath Patrick Geddes saw the error in that claim. Like Kropotkin he used his insight to develop a political creed.

Geddes had been a student of Thomas Henry Huxley at the Normal School of Science in South Kensington. He disliked life in the big city and in 1877 rejected a post at University College London in favour of a job at the marine station at Roscoff, in Brittany. There he worked on an obscure green worm. He found that it lost its colour if kept for long periods in the dark and did not get it back when returned to daylight.

The young Scot moved on to study green sea-anemones and discovered that they behaved in just the same way. Like worms, they had been assumed to be halfway between the vegetable and animal kingdoms. He suggested instead that such green creatures are a consortium of two elements, a plant and an animal. The larger party gives shelter and food to its resident, who pays back the debt as it soaks up the sun's energy to make carbon compounds. Without bright light, the crew abandons ship and leaves the worm or anemone to fend for itself, but in the laboratory at least each member of the coalition can cope on its own.

Patrick Geddes knew Kropotkin and was happy to use his own discoveries as an alibi for their shared social views. Cooperation, he was sure, was in our nature and anemones proved 'the bosh of saying that War is necessary to mankind and inherent in him'.

The anemone and its coral relatives depend, on a single-celled beast called a dinoflagellate (from the Greek for 'spinning whip' after the propeller used by many kinds to swim). The creatures have an illustrious pedigree, for they have a lot in common with the marine masons who built the stromatolites. Those ancient architects stole energy from sunshine, pulled carbon dioxide from the skies and combined it with calcium to make solid stone. Their efforts piled up much of the modern landscape.

After a billion years of labour they faded from view. The next generation of reefs was built by sea-shells, stony algae and the first (and helper-free) corals. The polyps themselves did not lay down much rock until their arrangement with an industrial partner began. The dinoflagellate symbionts of modern hard corals leave their signature in the shape of an unusual form of nitrogen within the tiny protein fraction found in their massive productions. The same chemical clue is first seen in fossil material from 240 million years ago, a hint that the partnership began about then. Once established, the business flourished. It crafted the atolls and barriers of today, which retain through their internal assistants a direct link with the lumpish objects built at the dawn of time.

A polyp has a million or so helpers in each square centimetre of flesh. Most are held in a thin layer on the upper surface. Sometimes they pass to the next generation via the host's eggs, but in many cases a newborn animal has to pick up its own internal cargo from the water.

Life as a paying guest is, from a dinoflagellate's venerable point of view, rather a novel escapade. Such creatures have flourished as independents for far longer than they have been members of a consortium. Many modern kinds remain aloof from any such alliance. Free-living versions are still important in the ocean's economy for the skeletons of some species form beds of ooze which can harden into limestone. Others are parasites of fish and crabs, while some generate the red marine tides that kill fish and, now and again, people. Captain Cook had a nasty experience off New Caledonia. He dined on a fish 'with a large ugly head' and became violently ill. He had a strange problem: 'Nor could I distinguish between light and heavy bodies, a quart pot full of Water and a feather being the same in my hand.' In time, 'A good sweat had a happy effect' and he recovered. Cook was afflicted by ciguatera, an illness caused by a dinoflagellate that eats dead coral. Its poison is concentrated when eaten by predators. The explorer was lucky, for some people suffer for months and a few die.

Important as they might be, such creatures are not well known to science. It takes a dozen of the largest dinoflagellates to stretch an inch and most species are hundreds of times smaller even than that. Each has two whips, and while one rows the other steers. Some kinds glow at night, helping to give tropical oceans their eerie light as boats or fish move through the water. Many have huge amounts of DNA – a hundred times as much as any human cell. Their genes are quite different from our own or those of corals, and they even have their own version of the genetic code. Their chromosomes, too, have a strange and unique design, perhaps because they contain such vast quantities of DNA.

The sex lives of the dinoflagellates are little understood, although some alternate – as we do – between phases with single, and others with double copies of their genes. They pick up genes from other sources and as a result bear bits of DNA, from a diversity of other single-celled beings. Once, all the symbiotic forms were assumed to belong to the same species, but the double helix reveals a hundred hidden kinds and the real number is bound to be more. Some are tied to a particular coral host, while others are promiscuous. The extent of divergence in their DNA shows that the various strains are so distinct that they must have been separate for millions of years.

Marine dinoflagellates have eclectic tastes, forming associations with clams and sponges as well as with corals, anemones and seaside worms. As a result, they power many of the engines that put down stone. Even so, plenty of corals manage without them. The red Mediterranean kind grows – slowly – by its own efforts, while those in the black deeps of the North Atlantic have no use for sunlight. Some of the hard corals of the equatorial oceans can also make a solid skeleton without the benefit of a symbiont. Most of the tropical species, though, depend on their consorts to make an impression against the waves. Marx would be dismayed to see that his marine builders are not noble workers in a common cause, but depend instead on the labour of others.

Those helpers provide impressive amounts of energy. A polyp so blessed makes, on the average, three times more coral in the light than in the dark – and in ideal conditions its residents can push up the work of production a hundredfold. Darwin tells the tale of a ship in the Persian Gulf whose copper bottom was encrusted in the course of twenty months with a layer of stone two feet thick: a feat that, unknown to him, turned on the efforts of a third party. The polyps do pick up a few nutrients as they eat tiny particles that float in the water, but their solar panels provide far more. They lay down carbon at almost twice the rate of a rainforest, making shallow-water tropical reefs the most productive natural places on the planet.

The economics of the coral factory resemble those of the industries of Merseyside or Denmark for the dinoflagellate soaks up its partner's wastes in the form of ammonia, carbon dioxide and hydrogen ions, while the host uses molecules gained from its partner. The symbionts hand over nine-tenths of their output in the form of sugars, carbohydrates and the building blocks of proteins. That raw material is transformed and, in part, returned. They pay a substantial rent, for they divide at a far lower rate inside a foreign cell than they do when free. As in industrial symbiosis, the sums are finely balanced. *Hydra* has green residents of its own, but in the dark they slow its growth. What is more, when food is abundant polyps without such visitors grow better than those that stay green. Only in daylight and when times are hard do its tenants pay their way, for then they make a useful sugar that gives their bearers an advantage.

Atolls and their relatives show how delicate the symbiotic contract can be. They seem a model of mutual support for the coral animal provides shelter and food while its consort soaks up sunlight. In fact, their submarine union is always on the edge, with the guest in constant danger of forced expulsion or voluntary exile, and its host of a solitary existence that may be bad for its health. The arrangement thrives when times are good, but may split up if they get nasty. Such uncertainty has a lesson for those who like to read political messages into Nature.

In Edinburgh, Patrick Geddes, the discoverer of the anemone symbiosis, became known to his colleagues as the 'Professor of Things in General'. He was the architect of the city's Camera Obscura, designer of the street plan of Tel Aviv and founder of the Sociological Society. In time he wearied of the Scottish climate and moved to Montpellier where – in a prelude to the mass migration sparked off by Ryanair a century later – he set up a Collège des Ecossais in a chateau bought with the proceeds of his many books. There Geddes worked hard to revive the 'Auld Alliance' between Scotland and France and preached his 'three S's' – Sympathy for the environment, Synthesis of the arts and the sciences and Synergy or collaborative action – to an audience of advanced young men and women.

The students particularly admired his views on the evolution of sex, which was, said Geddes in his tolerant Victorian way, a marvellous example of biological cooperation. Woman was complementary, rather than submissive, to man. Her nurturing tendency shaped the social environment, for she was vegetative and conservative and stored energy to be released by her radical and animalistic partner. The Scots philosopher did not, however, wish to go too far and opposed votes for complementaries on the grounds that 'what was decided among prehistoric protozoa cannot be annulled by acts of parliament'. Sex was mutual sacrifice – symbiosis in its purest form.

Since his day there has been a rethink about both practices. For each, the boundary between cooperation and conflict has become murky indeed. Modern views of reproduction are rather less magnanimous than were those of the Professor of Things in General. Advances in biology mean that we have been forced to add a potent extra 'S', Self-interest, to Patrick Geddes' erotic catechism. Two parties are of course involved in any relationship and an instant of cooperation is hard to avoid, but in many ways the strategy of each is quite different. The divergent interests of males and females can cause evolution to proceed at enormous speed. Sex is full of antagonism and symbiosis is much the same.

Many sexual disputes are about investment; about who brings in the cash and who spends it. Females tend to put more into childcare than do males, for eggs are larger than sperm, the embryo soaks up energy and babies must be nurtured until they are old enough to look after themselves. All this limits the number of offspring she can have, while her mate is less constrained. As a result, he may have a chance to slide off to find a new consort, while she must try to extract whatever she can while he is still around.

Males and females need each other, but each has an agenda of their own. Both parties must check that the other is not about to cheat. Their mutual suspicion expresses itself in a multitude of ways. Sperm outnumber eggs, but eggs evolve molecular locks that ensure that only the best get in. Some female insects soak up a male's ejaculate to give themselves a free meal, while her partner responds with poisons in his vital fluid that force her to lay eggs. Pregnant mice absorb their own embryos in order to have sex with a novel paramour, while male lions eat the cubs of a new and already pregnant mate. All this can lead to a series of signals in which one party (often the male) affirms his worth, while the other indicates her readiness, or otherwise, to cooperate.

Symbiosis, too, is a dialogue in which both participants must stay alert. The decision to start a relationship is in part the dinoflagellate's own, for it can sense the presence of a healthy polyp and swim towards its chemical signal. When a young polyp picks up its first partner it alters the activity of several dozen genes, as a hint that negotiations have begun. Several belong to the system of programmed cell death that plays an important role in development. The new arrival persuades the cells within which it sits not to kill themselves as instructed but to survive and to nurture their new visitor to their joint benefit.

Negotiations may become bitter. Quite often, the affair comes to an end. The assistant produces a signal that disables the host's defences, but when its ability to pay the rent drops too far, it is attacked and the cells in which it lives commit suicide. When the polyp does not meet its obligations as a landlord the tenants make nitric oxide, a toxic chemical. This allows them to make a break for freedom while their host is distracted by a desperate attempt to stay alive.

Like males and females, hosts and symbionts are in constant dialogue and either one may abandon an alliance as conditions change (although, as in many marriages, the weaker element is sometimes trapped in a liaison from which it cannot escape). Patrick Geddes would be depressed by that triumph of strife over synergy. His vision of biology as altruism has given way to a more realistic model based on selfishness and suspicion.

The plague of bleaching that now and again spreads across the oceans is a symptom of how the relationship strays close to collapse. In the summers of 1997 and 1998, the hottest for a thousand years, one-sixth of the planet's corals lost their colour as their helpers moved out. 2006 was almos as bad. The funeral bell was tolled by the Greens but many recovered within a few months. The symbionts were back in residence, but in a new guise. One type had been, before the disaster, rather rare. Afterwards it was found in many places. Like rosebay willowherb, the plant that colonised the ruins of London after the Blitz, it could deal with harsh conditions and flourished even within an overheated host. After 1945, as London life came back to normal, the weeds were lost. Their marine equivalents, too, were replaced with the original forms, back from exile in the ocean where they swam free until the heatwave was over. Bleaching was a move in an endless battle, an artful trick by landlords to expel tenants that could no longer pay the bills and to replace them with others more able to cope.

Across the globe, corals face many ailments, from white plague to pink line syndrome to black band disease. Most are a symptom of a breakdown in the relationship with dinoflagellates as life gets tough. Throughout history, reefs have always been liable to calamity and their symbiosis is in part to blame. After the disaster that killed off the dinosaurs, polyps with such assistants survived less well than did those without. Perhaps the loss of their associates under stress led to the collapse of the whole system.

Many corals welcome a variety of dinoflagellates and most of their helpers are happy to live in several kinds of coral. As a result, wife-swapping of the kind seen after the 1990s heatwave is rife. DNA exposes an uneasy marital history for it shows that the family trees of the polyps and their consorts are quite unalike. Each has switched its allegiance, sometimes several times, since the relationship began.

Squirrels and birds both make nests, but they are not close kin. In the same way, reef-builders come from several lineages. Rocky corals are not members of a single clan but are a diverse bunch of distant relatives, each of which has taken up masonry in quite recent times and has relations who do not bother with it. The habit began when some soft-bodied cnidarians that used dinoflagellates to make food from sunlight – anemones, as we call them – discovered a new way of life, based on stone. In those happy times of long ago, in sunny oceans filled with calcium carbonate, the assistants helped their hosts lay down solid material rather than pumping its constituents back into the sea. In time, a pliant polyp evolved a solid skeleton.

For a long era, the arrangement thrived. Then came a change. Around 120 million years ago a shift in climate and an increase in the acidity of the sea made it harder to put down masonry and altered the economic balance. Many of the builders went out of business as their labour force could no longer pay for their upkeep. Some were driven to extinction, but others fired their workers, reverted to the flexible existence of their ancestors and went back to life as a sea-anemone. As a result, a certain group of anemones today is more related to solid corals than it is to any other anemone.

In mutualism, infidelity is universal. Even lichens, the epitome of cooperation, have had several upheavals. Just as in the polyps, the family trees of the fungi and their assistants look quite different. Dozens of fungi that once lived as lichens have abandoned their relationship and the other party has returned the compliment. Even the mould that makes penicillin was once part of a lichen, but for some reason it has gone back to a solitary existence.

In evolutionary terms the dialogue between corals or fungi and their dinoflagellate helpers is rather new. They, together with almost all other creatures, are tied into a system of mutual aid – or sullen antipathy – far older and more profound than anything imagined by Kropotkin.

All cells more complex than bacteria are built upon such unions. The new map of the DNA of polyps, plants and people reveals a series of ancient mergers, in which genes of quite different origin were shuffled together to make the cells of today. As a result, every cell is a genetic multinational made up of once independent elements.

Patrick Geddes first saw symbiosis in an animal that was in part a plant. As he pointed out, 'By leaves we live': all of us depend on those helpful beings for every animal eats shoots and leaves, or devours other creatures that feed upon them. Their green pigment, chlorophyll, is contained within structures called chloroplasts. It puts the production line that breaks down foodstuffs into reverse. Photosynthesis, as the process is called, takes the energy from sunlight into the cell's chemical factory. Plants soak up atmospheric carbon dioxide and break down water into oxygen, hydrogen ions and a stream of charged particles. The oxygen is released and the other products link themselves to carbon to build new molecules.

A plant cell contains many chloroplasts, each with its own complement of the double helix. Such structures – like the builders of the first reefs – helped to transform the Earth from a fetid charnel house into the planet of today and do their best to keep it that way. They have an origin quite separate from that of the cells that they inhabit.

Their genetics was first studied in a single-celled green plant that lives in fresh water. Now and again a mutation may damage its chlorophyll and produce a colourless individual. Crosses between green and white lines showed that the mutation was inherited in an odd fashion, for it passed to the next generation not through both parents, but through the mother alone. Unlike the male, who transmits only genes in the nucleus of his sperm cells, females also hand on cytoplasm – the main mass of the egg – to their progeny. The chloroplast genes, their pattern of inheritance showed, must be in the cytoplasm rather than in the nucleus. Further crosses revealed that they are arranged not in long strings like those of the nucleus itself, but as a small closed circle.

The arrangement hints at an unexpected history. In 1905 the Russian biologist Constantine Mereschowsky came up with the radical notion that chloroplasts had once been distinct, 'little workers, green slaves'. They had, he claimed, been hijacked by a more powerful entity in the distant past.

The notion sounded absurd, but is quite correct. Chloroplasts have an odd design. Each is enclosed within a membrane and has a series of internal barriers, rather like the plates of a car battery, upon which light is trapped. That arrangement, their closed circle of DNA and the details of their genetic code are very like those of bacteria. Long ago, a bacterium entered a cell and has done service as a solar panel ever since.

DNA reveals an unexpected tie between Nature's gardens above and below the sea, for the chloroplast shares many of the genes of the coral symbiont. Plants and corals live in very different places and use the Sun in different ways but the similarity of the DNA of their internal assistants shows that each of their symbionts shares the same ancestor. A very similar molecular signature is found in lichens and in dinoflagellates themselves. The symbioses of all these quite distinct solar-powered creatures spring from a common root.

The veritable Adam (or Eve) of all today's billions of solar harvesters entered a primitive cell just once. On that historic day, around a thousand million years ago, a simple organism that looked rather like a modern amoeba, with a similar ability to swallow up small objects, engulfed a bacterium able to soak up the Sun's energy. Its enhanced self thrived, divided and became the forefather of all today's sun-worshippers, be they the architects of Cocos–Keeling Island, the builders of the Amazon rainforest or the lichens of an English country churchyard.

The genes of the helpful light-gatherers, on land and sea, are close to those of a modern bacterial group called the cyanobacteria. Their tie with the reefs is ancient indeed, for three billion years ago their ancestors made the stromatolites, the earliest fossil reefs of all.

The cyanobacteria who built the first walls of marine stone are still busy at a diversity of tasks today. Many find their homes in wet places and, as a result, give earth its distinctive smell. Some are green (a health food is based on a species called *Spirulina*) but some prefer other colours – the Red Sea is named after the red blooms of a cyanobacterium and flamingos are pink because they eat another kind. In today's tropical seas and rice paddies such cells form blooms that pull in nitrogen and fertilise the otherwise sterile waters. Others cluster around the roots of peas and beans to supply them with the same fertiliser.

The emergence of the cyanobacteria, with their ability to make oxygen, was the most important event in the history of life. Before that, biological commerce depended on sulphur, carbon, iron and hydrogen. Their efforts increased the flow of energy through ecosystems by thousands of times and sparked off an explosion of evolution.

Our planet has breathed free for half its existence. Its ability to do so was soon improved by the many species that took up those simple bacteria and put them to work for their own ends. Dinoflagellates and green plants were but two of many creatures to enter into an arrangement with cyanobacteria.

Most of those ancient liaisons have a more complicated history than that of the green plant. Many of the sunbathers obtained their helper with the aid of a third party, for they sucked in a cell that had already engulfed a solar panel of its own.

Today's dinoflagellates show how complicated the transactions have been. In some species the internal labourer is bounded not just by a single cell membrane – that of the original bacterial slave itself – but by two, one inside the other. Others have three or four membranes, proof that some swallowed their helpers directly while others engulfed them at second, third or even fourth hand. Masters became servants more than once, and when a dinoflagellate is taken up by a coral another step is added to the industrial chain. For most sun-worshippers, life is a Russian doll, with form within form in a hierarchy of dependence. Mereschowsky, the green slave man, would be delighted.

After any merger a period of negotiation is inevitable. Some symbionts – those within a coral included – keep most of their own machinery and can escape should they wish to. For others, fusion is followed by exploitation and bondage. DNA reveals a history of ruthless asset stripping of their helper by the plants and animals that made use of solar technology.

Almost all the chloroplast's production machinery has been transferred to its partner's nucleus, with fewer than two hundred genes left in the green slave itself (compared to several thousand in its free-living relatives). A leaf is full of verdant skeletons. Their hijacked machinery has been put to good use. A fifth of the genes in a green plant's nucleus have been shipped in from its chloroplasts. Some still help to harvest light, but they also play a part in cell division and even in resistance to disease. The host, in return, uses its own DNA to deliver proteins needed by its outworker.

The trade still goes on. Rice has transferred chloroplast genes to the nucleus within the past million years, for in its close relatives the grasses they remain in their original location. The process can even be seen in the laboratory. A chemical label attached to the chloroplast DNA of tobacco shows that one cell in five million gains a new nuclear gene from its tiny partner within a few days. As a result, one plant in a few thousand – a dozen or so in any field – has hijacked some DNA from its erstwhile serf.

Material always flows from symbiont to host and not the other way. If a chloroplast – just one of many such labourers in the plant's cell – bursts, the factory in which it works survives for long enough for some of the released DNA to move to the nucleus. If the nucleus itself – the control room – explodes, the cell dies at once. That one-way street to decline means that the green serfs can no longer make a life of their own. Instead, they generate food for their owners and oxygen for the world, with no chance of a break for freedom.

Even chloroplasts, subservient as they are, arrived rather late in biological history, for most creatures manage without them. Symbiosis itself began long before the first green slave, or the first coral.

Almost all cells contain mitochondria, the structures that use oxygen to burn food and to pump out energetic molecules together with dangerous pollution. They are active indeed, for each of us produces almost our own body weight of their products each day.

Like chloroplasts, mitochondria were once bacteria and, like them, they still take in raw material from their surroundings – the cell – and break it down. The structures retain some autonomy, for each divides independently of the nucleus and has DNA of its own, arranged in a circle. In the laboratory at least, they can move between cells. When human cells with damaged mitochondria are cultured together with their normal cousins the functional versions may move across to aid their crippled kin. Many inborn conditions are due to errors in mitochondrial DNA and many more are caused by mutations in nuclear genes that came from the same source long ago.

Those useful structures entered cells around two billion years ago. Unlike chloroplasts, all of which look very similar in plants from bananas to seaweeds, mitochondria have evolved into vast variety. Our own and those of the cnidarians are compact, with almost no spare DNA, but others are gothic in their complexity. In the human version just a few genes involved in protein synthesis, together with those for thirteen proteins that burn oxygen, are left. A thousand and more genes belonging to the ancestral bacterium are scattered through our nuclear DNA. In some simple beings the mitochondria are left with no more than three genes of their own. Even so, all such structures are similar enough to suggest that the invasion took place just once.

As is the case for corals and their partners the interests of each party differ, despite a veneer of cooperation. The disputes are more subtle than upon the reef, but are just as lethal. The symptoms of age are side-effects of their apparently selfless task, as they generate reactive oxygen. That is bad news for those who bear them, but by the time senility strikes the mitochondrion's own DNA has been passed to the next generation through its host's egg. Its decline is hence of no interest to the energetic little partner.

Other elements of cells also emerged as part of a grand alliance. The whips that propel them or move mucus over their surface were once bacteria. Part of the mechanism which helps cells to divide might also have come from outside. Kropotkin's mechanism has been appealed to, or speculated about, to explain many other properties of life. A sperm is swallowed by an egg in a process that looks rather like the ancient fusions that marked the earliest days of the cell – a model of sex as symbiosis close to that of Geddes' own scheme.

The most unexpected candidate for ancient independence is the cell headquarters itself. Polyps, plants, plovers and people are proud to call themselves eukaryotes. Their cells have a nucleus, embedded in cytoplasm. The idea that the structure might be an import is not new, for Thomas Henry Huxley made rather a fool of himself after the *Challenger* expedition when he claimed that the sea floor was covered with an organic jelly. That primordial protoplasm looked just like a sheet of cells, each without a nucleus, which, Huxley suggested, had invaded later (it was in fact a chemical precipitate made by exposing gypsum to laboratory alcohol).

Huxley's claim was absurd, but molecular biology gives new life to the idea of a separate origin for the cell's control centre. The first creatures to be blessed with a nucleus evolved a billion and a half years ago, about halfway through the history of existence. Bacteria, which lack such things, are far older. They had split long before from another distinct kingdom, the archaea, who also manage without a separate centre of operations. Nowadays, archaea are found in oceanic hot vents and other exotic places.

Bacteria and archaea run their lives in a single space. In effect, they are companies with the head office on the shop floor. In contrast, eukaryotes, with most of their DNA held within the nucleus, keep the information end of the business separate from the vulgar process of manufacture. The management – DNA and the enzymes that copy and read it – stays within an inner compartment while the machinery is outside, in the body of the cell.

A small messenger molecule flows out of the planning department to the factory floor and tells the enzymes what to do. The system is far more efficient than that of archaea or bacteria.

Human, plant and coral DNA is, in general, more similar to that of archaea than to that of bacteria. Even so, some of our genes look more like those of the latter group. The DNA hints at a clear division of labour. Archaeal genes run the information side of the enterprise – the proteins needed to manage DNA – while those with a bacterial identity spend their time on the production line itself. Their job is to generate energy, keep the place clean and make the products exported by the biological factory.

A central part of the cellular economy has, it seems, been shipped in from another supplier. All eukaryotes began with a merger between a bacterium and an archaean. As the years passed, the ancient archaean became what is now the nucleus and its bacterial partner evolved into the cytoplasm. That exchange was, like that on the reefs, based on a trade in raw materials. In the oxygen-free atmosphere of long ago the archaean gained hydrogen from its partner and provided methane – a compound of hydrogen and carbon – in return. The chemistry of the two processes did not mix and the reactors stayed in separate buildings, screened off by a wall, the nuclear membrane. As oxygen filled the skies the archaean could no longer get hold of hydrogen, which was now combined into water. It began to depend more and more on its bacterial partner until at last the two became a single cell.

Symbiosis marks each stage in evolution, but the notion of mutual aid, of joint effort to a common end, has been superseded by a sterner view: that such arrangements began with simple exploitation. Disease, parasitism and cannibalism have been around since life began. Some of those ancient conflicts have matured – or decayed – into what we choose to see as amity. The evidence lies in the infections of today.

Sometimes a disease kills us and sometimes we beat it off, but quite often we reach an uneasy state of chronic illness that may, in time, become quite acceptable. The invaders evolve to become milder, keeping their reluctant hosts alive for long enough for their enemies to find another home. The malady may even settle into an arrangement not unlike that of certain symbionts. Two billion people across the globe have tuberculosis, but most have no idea of their condition and a mere three million die each year. In that disease, as in many chronic conditions, the pathogens are walled away in some remote section of the body to reduce the damage, in a pattern that also applies to certain symbioses.

Gut bacteria, too, have an ambiguous relationship with their proprietors. They are in constant conversation with the local intestine and evolve to adapt to a new home. Such genetic negotiations mean that the bacteria in the bowels of identical twins living apart are more similar than are those of husbands and wives in the same household. The genes involved act as a flag of truce that persuades the host to accept the visitor as harmless – but they are almost identical to others used by harmful bacteria to evade the body's defences. As any traveller knows, a visit to a distant place with a bacterial identity of its own means a sudden shift from cooperation to conflict, with dire effects on the unfortunate tourist's gastric state.

The double helix suggests that many symbiotic partnerships have had a parasitic past. Many pathogens have evolved tricks that overcome a series of security mechanisms on the cell surface to enter the sanctum itself. The agents of dysentery, syphilis and the plague all suck cells dry. They have a variety of genes that act as skeleton keys, mimic the host's cues of identity and disable its protective mechanisms. Malign as their effects might be, they look remarkably like genes found on the surface of the coral symbiont.

Many symbionts have relatives that still cleave to a parasitic past. The closest kin to mitochondria are the rickettsias, tiny bacteria too small to see under the microscope. They cause typhus and trench fever and can reproduce only within an animal cell. Rickettsial DNA is similar enough to that of the mitochondrion to suggest that today's universal symbiont itself started as an aggressive pathogen. In the same way, the ancestors of the cell's system of beating whips upon the surface look rather like the modern agent of syphilis.

Corals are a perfect illustration of the intimate tie between parasitism and symbiosis. Some polyps pick up their assistants while young from dinoflagellates in the water, some pass them on direct to their progeny when they split or lay eggs, while others use both strategies. A simple experiment shows how apparent aid can soon change into overt enmity.

Polyps kept in the laboratory were persuaded to pass on their auxiliaries directly for several generations, with no chance to import them from outside. Others were given a supply of free-living dinoflagellates but were not allowed to transmit them down the generations. From the symbiont's point of view, the regimes are quite distinct. A line that passes them on through the cell gives the assistant no chance to indulge in selfish behaviour at its host's expense, for to do so would reduce its own hope of survival. In those circumstances the associates stay subdued and helpful. With free exchange through the water, competition sets in and the visitors can set their own agenda. Soon, the relationship looks more like parasitism than aid. The dinoflagellates copy themselves faster, use more of the polyp's resources and slow its growth. Those newly emboldened beasts leave for a better life at any provocation, while their vertically transmitted brethren stay as helpers through thick and thin, for they cannot pass on their own DNA until the polyp does the job for them.

The genes switched on by a young polyp after the entry of its first symbiont persuade its cells – and the new visitors within – to survive for longer than normal. Parasites use the same strategy for selfish ends. An infected host tries to destroy its unwelcome guests with the forced suicide of diseased cells. They, in turn, disable plans for self-destruction and force the host cells to stay alive.

Some parasites are already on the way to symbiosis. The single-celled creature *Toxoplasma* is carried by half the human population. It sits within the immune system for life. That devious relative of the dinoflagellates allows most of the mammals that carry it to stay healthy because it can reproduce only in the guts of cats – which eat only live and vigorous prey. Infected cats become ill as they excrete billions of cysts, but not until a man or woman is weakened by AIDS or by chemotherapy does the mean-spirited *Toxoplasma* cause any damage.

The road from enemy to friend is a two-way street. Malaria parasites – themselves relatives of *Toxoplasma* – contain a strange object that helps them to invade blood cells. It has DNA of its own which, quite unexpectedly, resembles that of a chloroplast and shows that the disease that kills a million a year is caused by an organism that once lived free. Many of the malaria parasite's genes are close to those of dinoflagellates. Its ancestors, like those of an oak tree, were green, but in time it became a pallid freeloader.

Toxoplasma and the agent of malaria move from cell to cell by hijacking a protein on the surface, and the molecular receptor involved resembles that used by corals in their negotiations with dinoflagellates. Their ancestry also means that each still shares a few genes with plants. There lies their Achilles' heel. Certain weed-killers attack chloroplasts and an ingenious doctor suggested that they might work against malaria. A herbicide is already being used to treat children with the illness. A new era of drugs might be set to flower as man gets his own back on a helpful creature that turned to selfishness.

From tropical diseases to lagoon islands, the science of symbiosis has moved on since Beatrix Potter. Her placid notions of mutual aid now look tired at best. Kropotkin himself went even further in insisting on the long-term power of mutualism as an evolutionary force. Cooperation would continue until it reached perfect harmony, to the benefit of all involved. He ascribed to the world of life a most useful talent: the ability, like Mrs Tiggy-Winkle herself, to plan ahead in the interests of its descendants. The naturalist on

Cook's second voyage fell into the same trap when he asked about atolls: '. . . how it comes that the *Madrepores* form such circular or oval ridges of rocks?; it seems to me that they do it by instinct, to shelter themselves the better against the Impetuousity & constance of the SW winds; so that within the ridge there is allways a fine calm Bason, where they feel nothing of the Effects of the most blowing weather.' Like Karl Marx he admired the perspicacity of the polyps, their ability to work together to a common, albeit distant, end.

The bold traveller, the Russian prince and his philosophical countryman each ask too much of the foresight of animals. Men build harbours on such a principle but corals cannot plan for the future, whatever benefit that might bring. Economics may help to understand evolution, but our own financial equations have moved far beyond those of the animal world. Long ago they entered a domain about which biology can say nothing.

Marx, in *Das Kapital*, insisted that cash is the product of an economy and not the economy itself. Geddes agreed: 'Some people have the strange idea that they live by money. They think energy is generated by the circulation of coins . . . [but] we live not by the jingling of our coins, but by the fullness of our harvests.' Money is the memory of past bargains. A deposit – or an overdraft – in the financial memory bank is a token of an obligation fulfilled or still unsettled. It adds virtual reality to the realm of goods and labour, for it separates the act of purchase from that of sale. All human symbioses, industrial or otherwise, involve some form of that ancient habit 'you scratch my back and I'll scratch yours'. Cash measures who has scratched whom, how hard and how often.

Its biological equivalent also demands a certain power of recall, for a crocodile must remember that plovers and not pigeons clean its teeth. In the same way, the cell signals involved in symbiosis reassure each party that the other has paid the rent. Unlike global capitalism, though, symbiosis thinks only in the short term. It has no concept of assets held in hypothetical form, or of benefits long deferred.

Most attempts to read economics into Nature ignore that crucial contrast. The physicist James Lovelock noted that the atmosphere of Mars, an inert mix based on carbon dioxide, was just like that of our own planet before the reef-builders got busy. He suggested that the ancient biosphere – land, sea and air – was a grand symbiosis that transformed a hostile atmosphere into the friendly skies of today. The Earth was not a 'demented spaceship, forever travelling, driverless and purposeless' but an organism evolved towards harmony, under the influence of an Earth Mother 'with faculties and powers far beyond those of its constituent parts'. In an unexpected tie to the atolls, her name, Gaia, was suggested by his neighbour William Golding, author of *Lord of the Flies*.

That notion was popular in the 1960s (it was no coincidence that the gang that kidnapped Patty Hearst called itself the Symbionese Liberation Army) but fails the test of accountancy. Nobody, altruistic though he might be, will work at a job in which payday does not arrive until long after his own demise. A reward too long delayed is no reward at all and the helpful bacteria that first allowed cells to make oxygen had no notion that their labours would, billions of years on, give rise to creatures that depend on it. Lovelock's theory resembles that of 'intelligent design': the denial of evolution on the grounds that complex structures could not emerge without forethought. To a Darwinian, life has no strategy. It evolved through a series of short-term tactics, and the symbioses that make the modern world emerged in the same messy, expedient and arbitrary way. Corals, oaks and human cells show that we do live on a purposeless spaceship. Fortunately, for most of the time we do not notice.

Sadly for Patrick Geddes and Karl Marx, the Slavic experiment in mutualism that followed the Russian Revolution failed. Cash did not, as predicted, disappear and in Russia roubles and inequality still flourish. Like Russia, the reefs – and the cell – depend not on mutual aid but on greed and mutual exploitation. Kropotkin would be disappointed to learn as much, but his own followers soon sank into the same habits. Quarrels between thinkers became wars among doers and, as ever, those on the left were the first to fall out. In the early days, the adherents of Marx and Kropotkin saw eye to eye, but at a congress of the International Workingmen's Association they disagreed about whether to pursue immediate revolution or to work towards political change.

The anarchists were expelled. They split again and again. When persuasion failed some turned to the ideas of Nietzsche. They assassinated the presidents of Italy, of the United States and of France in the 'propaganda by deed' that gave a notable impetus to the bloodbaths of the last century. A few stayed true to their gentler roots – but nowadays, alas, such people are on the fringes of politics, sidelined by the iron rules of greed that rule the globe.

Biology will not help them, for scientists have nothing to add to philosophy apart from facts. As their work goes on new facts emerge. Some may be uncomfortable, but philosophers will no doubt continue to appeal, like Kropotkin, Malthus, Nietzsche, Geddes, Marx, Lovelock and more, only to those that remain in happy symbiosis with their own beliefs.

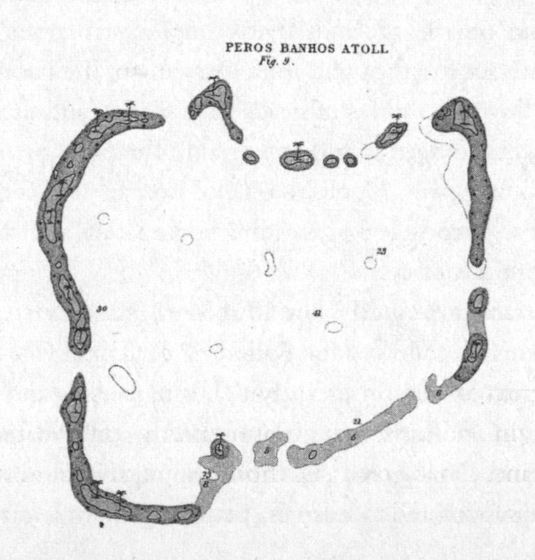

CHAPTER IV

THE EMPIRE OF CHAOS

In science the notion of cause and effect can be hard to escape. The Bostonian cleric Dr Price regarded the earthquake that struck Massachusetts in 1755 as the direct result of the introduction of lightning conductors. Those 'iron points invented by the sagacious Dr Franklin' impeded a divine instrument for the punishment of sinners and the disaster was sent as a result. The fit was precise: 'In Boston are more erected than elsewhere in New England and Boston seems to be more dreadfully shaken. O! there is no getting out of the mighty hand of God.' His grim philosophy is still alive, for after the Indian Ocean tsunami of 26 December 2004 a Saudi religious authority noted that the wave had struck hardest in the resort of Phuket, where 'corrupt people from all over the world come to commit fornication and sexual perversion'.

The place and time of the next tremor cannot in fact be divined from the distribution of lightning rods or perverts, but even modern seismology, with all its instruments, is remarkably bad at saying just when and where it will happen. Its failure reflects a great change in the way that science sees the world.

To Darwin and his fellows Nature was a complicated but efficient machine. Biology was, in those happy days, almost as simple as the solar system, with animals, plants and people circling, like planets, in their ordained paths. Life followed laws that, once deciphered, would reveal a pattern that stretched from the calm of the English landscape to the vast diversity of the tropics. Every atoll, every forest and every mountain peak was filled with just as many species as it could support, each perfectly adapted to where it lived. Once Nature's mechanism was understood her operations would be as predictable as those of the heavens.

That comfortable philosophy had worked well for the planets and to an extent still does. The 8 June 2004 marked one of the infrequent returns of the Transit of Venus, the event that sent Captain Cook to the Pacific. Across the globe, people knew to the second when its disc would touch the edge of the Sun and planned their day to match (even if my own attempt to view the display was thwarted by the vagaries of the London weather). There is something marvellous in the ability to tell far in advance exactly when such an event will take place. In Tahiti the next Transit will reach its maximum at 3.28 p.m. on 5 June 2012, a statement that, in its certainty, proves what a harmonious and predictable place the universe must be.

In 1837, on his long ride through central Chile, Darwin had an experience that caused him to doubt that reassuring view. The city of Concepción, not far from his camp, was demolished by a huge tremor. As he wrote in *The Voyage of the Beagle*: 'A bad earthquake at once destroys our oldest associations: the Earth, the very emblem of solidity, has moved beneath our feet like a thin crust over a fluid; one second of time has created in the mind a strange idea of insecurity, which hours of reflection would not have produced.'

I know just how he felt. At noon on 13 August 1978 as I relaxed with the Sunday newspapers on the sunny campus of the University of California at Santa Barbara I was startled by a series of discordant peals made by the college's carillon. As the bell-tower itself whipped back and forth half a million books fell off the shelves in the library. Filled with a certain sense of insecurity of my own I rushed into the Biology Department, to find my lab plunged into a shambles of broken bottles and escaped flies. A seismic shake at five on the Richter Scale – a modest affair, with dozens larger in California since then – had struck.

The bells helped fix the incident in my mind, but I have no memory of the many other tremors I lived through in my years in the Golden State for they were too small to notice. California, staid as it is, is always a-tremble and even Britain gets several shocks each year. Most can only be picked up with sensitive instruments, but much of Lincolnshire had a perceptible quiver in 2008 and London suffers the occasional seismic frisson (a 1580 event is mentioned in *Romeo and Juliet*).

Each year quakes release the same energy as a hundred thousand Hiroshima bombs. They do plenty of damage. The Sumatra Boxing Day disaster in 2004 generated a tsunami, which battered many reefs. Some were smothered in the murky water while others were smashed by the waves. The Krakatau explosion of 1883 was even worse, for it moved six-hundred-ton coral boulders hundreds of yards inland.

We have a good general idea where such cataclysms are liable to take place: in regions where a mass of oceanic rock slides beneath a continent, or on the edge of a plate where sheets of rock grind past each other. The epicentre of the Boxing Day event was thirty kilometres below the surface, near a point where three plates – the Indian, the Australian and the Burmese – come together. The Indian plate races north at five centimetres a year and is dragged below its Burmese equivalent as it travels. The slip took place over some two thousand kilometres. The ocean floor off the Thai resort of Phuket, that home of the corrupt, moved by twenty metres as the sheets of rock tore past each other. Elsewhere the land rose upwards by three times the height of a man.

The devastation caused by the tidal wave that followed came as a shock to the public, but in the great sweep of history such events are commonplace. Sumatra itself was only the fourth-biggest tremor in the past hundred years (the worst was in Chile in 1960 in a quake even bigger than the event that disquieted Darwin) and in geological terms a century is no more than an instant. The Lisbon calamity of 1755, with its hundred thousand dead, is mentioned in Voltaire's *Candide* ('It had been decided by the University of Coimbra that the burning of a few people alive by a slow fire and with great ceremony, is an infallible secret to hinder the Earth from quaking . . . The same day the Earth sustained a most violent concussion') and a century and a half earlier the coasts of the Bristol Channel had been struck by a tsunami: 'Mighty hilles of water tumbling over one another in such sort as if the greatest mountains in the world had overwhelmed the lowe villages.' That event was described in a pamphlet entitled *God's Warning to the People of England*. The Welsh, it appears, were beyond advice.

Historians have noted a few such episodes but better evidence is written in corals. On the main island of Hawaii 110,000 years ago a gigantic tsunami hurled fragments of coral three hundred metres above the waves. Smaller shocks have marked the shallows. They too can be used to recreate geological history. Darwin himself noted after his Chile experience the presence of 'beds of putrid mussel-shells, still adhering to the rocks; ten feet above high water mark: the inhabitants had formerly dived at lower-water spring-tides for these shells'. Now we know that after an earthquake the shoreline may leap upwards to settle again while the tension builds up once more. Reefs perish as they are thrust above the waves and remain moribund until the ground sinks once again beneath the sea. Their tale of life and death is recorded in the rock.

Cores bored into patches of coral off Sumatra, close to the epicentre of the Boxing Day event, show that the region has suffered several huge tremors over the last millennium. Two took place in the thirteenth century, with two more in the next couple of hundred years. There were large disturbances with thousands of victims in 1797, 1833 and 1861. They were followed by the tsunami of 2004. A closer look at the marine record shows a series of smaller shocks in 1728, 1746, 1797, 1833, 1861, 1876 and 1893.

The Indian Ocean is a risky place to live and its reefs show that on a scale of tens of thousands, of a thousand and of little more than a hundred years the actual disasters – be they stupendous, huge or no more than serious – arrive like acts of God, with no apparent pattern. That observation is bad news for shore-dwellers, but points towards a deep truth in Nature.

A billion dollars' worth of seismic instruments is scattered across California and Japan, with more around the Pacific, but not one disaster has been predicted with any accuracy. The surface of southern California is heaving hard and a quake as big as the giant event on the San Andreas fault three hundred years ago is on the way – but the best the seismologists can do is to estimate a seven in ten chance that, somewhere in the Los Angeles Basin, there will be a great tremor at some time within the next thirty years. We know more about our planet than did the Reverend Dr Price, but our knowledge has been of almost no help in saying just when its uneasy mass will strike back. Geology has had to come to terms with the unfortunate fact that the precise time and place of such incidents is impossible to foretell.

Uncertainty now holds sway in many other sciences that impinge upon the tropical seas. That disturbs scientists, who have long assumed that the universe is ruled by order, with a place for everything and most things in their place. It is not. Atolls and their inhabitants illustrate, better than anywhere else, the power of chance. The plants, animals and people who have the dubious privilege of living upon them gamble with the unsettled movements of the Earth, the seas and the winds. As they do, their fate defers to the deities of chaos.

Under the new mantle of doubt a set of laws of uncertainty has begun to emerge. Most biologists suffer from mathematics envy; either because (as in my case) they are too dim to cope, or because their science longs for the deep arithmetic truths that, they believe, underlie the atom, the solar system and the universe. Darwin shared the sentiment: 'I have deeply regretted that I did not proceed far enough at least to understand something of the great leading principles of mathematics; for men thus endowed seem to have an extra sense. But I do not believe that I should ever have succeeded beyond a very low grade.'

Charles Darwin was, without doubt, better engaged in leaping across tropical inlets with a pole or – as he sometimes did – playing a bassoon to earthworms to test their ability to sense vibration than in arithmetic speculation. He thought a great deal, but in contrast to the thinkers who still decorate biology, he also looked for facts. Mathematics has given a new insight into the facts gathered by biologists and geologists themselves. It finds that many of them follow a universal and unexpected rule.

Earthquakes, random as they might appear, show one important regularity. Charles Francis Richter, he of the Scale, noticed that their intensity is related to their frequency, with huge numbers of small tremors, plenty that throw books off shelves and a very few able to match Sumatra 2004. Quakes, in other words, follow a 'power law', a straight-line fit between how severe they are and how often they happen.

That maxim applies in fields far from geology, or even from science. Some of the words used in this chapter are common – *the*, *of* – while some – *empire* and *chaos* included – are rare, but that banal observation hides an unexpected truth. The British National Corpus, a collection of a hundred million words of written and spoken English, gives the big picture. *The*, for example, ranks top, *empire* at number 2701, *of* at two, and *chaos* at 5000. A graph of the rank of each word against the number of times it can be found in any solid chunk of text reveals a few common words, a fair number present in medium abundance and vast screeds that are almost never used. If the axes are set in powers of ten – 10, 100, 1000 and so on – the line between a word's frequency of use and the intervals at which it appears is, like that of the intensity of earthquakes against their prevalence, quite straight. In the tiny sample of English represented by this book, for example, *the* turns up in every fourteen words, *of* every twenty-five, and *empire* and *chaos* just once in every 4676 and 7226 words respectively.

The power law applies on many scales and in many places. Giant rivers have the same patterns of meander across a plain as do the raindrops falling on a window, while avalanches look like scaled-up versions of the sand that, every few moments, slips off the face of a dune. Wars fit, too, with small campaigns far more common than vast confrontations (the First and Second World Wars, the worst in human history, killed two-thirds of the victims of all the 315 conflicts since the defeat of Napoleon). Even bird-watchers get their kicks from the rule, with a few common-or-garden species matched with some unusual sightings and, for the real twitchers, lots of collectors' items ready to be boasted about.

From genetics to geology and from raindrops to rarities the power law unites phenomena that seem at first unrelated. Who, before Newton, realised that planets follow the rules that control how an apple falls to the ground, and who, until mathematics got in on the act, imagined that earthquakes and English literature have anything in common?

Such patterns emerge from a delicate balance between order and disorder. The shift between the two can be seen in any kitchen when the tap is turned on. An even flow is suddenly succeeded by a rush of surging water. The rate at which a liquid travels in a pipe when things go smoothly depends on a simple fit between the diameter of the pipe and the speed and stickiness of the fluid. If one of those variables is pushed beyond a certain limit the system breaks down and – for any liquid, from water to syrup – disorder, in the form of turbulence, takes over. Quite when it will prevail is impossible to predict, but the law is inexorable.

The battle of disorder with its opposite is joined in many places. Rivers build up sediment on the broad, smooth and slow outer curve of a bend and remove it from the inner bank, where the water runs faster and in a more broken manner, to build up an unstable set of meanders. On dunes and snowfields, gravity tries to flatten the landscape, but the fickle movements of the wind and the random build-up of grains or snowflakes work in the opposite direction until, with no advance notice, the sand slips or an avalanche sweeps away a skier. Even the solar system has chance built in, for with so many planets it becomes impossible to know what, over vast periods, will happen. The Moon itself wobbles eccentrically in the sky (Newton's failure to understand its pattern gave him, he admitted, a headache) and astral disaster is, sooner or later, more or less guaranteed: Mercury may collide with its neighbour or Earth itself be ejected from the solar club, which would at least put an end to the Transits of Venus.

Richter's seismic patterns emerge for the same reason. As the continental plates move against each other the rocks stick. Energy builds up and, all of a sudden, is released. The Sumatra tragedy was no more than a magnified version of what happens with the kitchen tap – first a slow but steady trickle as the plates slide past and then, with almost no hint of what is to come, a violent rush of rock.

Like tar, our planet's mantle is liquid. It oozes rather than flows. Heavy objects such as mountains, with or without an atoll on top, sink into the sticky mass but sometimes the surface, relieved of a burden, rises instead. Today's instruments can measure the dimple caused by an office block or the slow decline of a volcanic island as it sinks into the depths.

In 1491 the people of Östhammar in Sweden petitioned that

their town be moved closer to the shore as boats could no longer get in to the harbour. They believed that the water had leaked away through cracks in the bottom, but in fact their native land was rising. In 1731 Anders Celsius (better known for his temperature scale) cut a groove in a rock at the high water mark and was startled to find an annual increase in the gap between the line and the sea surface by almost a centimetre. A quick sum led him to suggest that his country had once been an island; he was right, but the idea was dismissed as heresy. The nation was on the rebound from the burden of a lost ice-cap.

Scandinavia bounces back by half a metre a century as it eases itself of the dead weight of its ancient glaciers. They went long ago, but the landscape has taken a long time to stretch its limbs. Sweden's slow heave shows just how fluid our planet must be. The mantle is a billion billion times more viscous than water – which makes glass, itself a creeping liquid, look like a mountain stream.

The Earth moves at a slow pace indeed and is remarkably sticky, but at the edge of a plate enormous energies are forced into a narrow channel. Their release manifests itself as many tiny tremors – or a huge disaster. When, where and how big a slip might result is quite impossible to tell.

The shift from order to disorder was the death knell for the physics of Isaac Newton. Everywhere the traditional and direct fit of cause with effect has been forced to give way. From taps to tsunamis the dread empire of chaos has been restored. The reefs themselves give powerful evidence of the random arrivals of seismic tremors but their peoples have many other reasons to fear the onset of anarchy, in the air they breathe, the waters that surround them and the hostile universe of life with which they struggle.

The flapping of Amazonian butterflies that leads – famously – to storms in Scotland is the universal image of how a tiny disturbance may cause a great upheaval. That tale is overstated, but the currents of the skies and the seas are, like those of the rocks beneath our feet, driven by the conflict of order with its opponent. As solar energy streams in, clouds, rivers and ocean currents act as shock absorbers. They damp out what would otherwise be fierce swings in temperature from place to place or season to season. As it streams from a tap, pours from the skies or spins in vast gyres in the ocean, water as liquid or vapour often falls into the grip of confusion.

We can tell, in general, what the climate is like in a particular place at a certain season, but in many parts of the planet we are not much better at predicting the weather than we are at earthquakes. In Britain tomorrow's forecast is often accurate, but that for a week ahead is little more than speculation. Other regions have a more stable pattern, but even they may suffer unexpected storms or calms.

Pacific islanders have always known that disaster, if it comes, is likely to strike in a particular season. For cyclones, the clouds and waves gave them notice a day or so ahead. Beyond that was guesswork.

A tempest once called for Dr Price's Universal Explanation: the gods were angry. Then came a search for more mundane reasons. After the Twin Towers outrage of 2001 many Indians blamed a persistent drought on the smoke caused by the American bombardment of Afghanistan. The United States was not to blame. The explanation lay instead in the physics of the atmosphere and of the oceans.

On my way to work each day I walk past a handsome Victorian house, a few doors from my own abode. It bears a plaque to George James Symons, 'The Father of British Rainfall'. In 1858 – a drought year, in which the Great Stink of the filthy Thames forced the House of Commons to drape its windows in carbolic-soaked cloth – he put a rain-gauge in his garden at Camden Square, in north London. He advertised for others to do the same and soon came to the attention of Admiral Robert

Fitzroy (once Captain of the *Beagle*, but by then at work on a system of storm warnings at the Board of Trade). Fitzroy offered him a job. Hundreds of people sent their figures to Symons, who collated the information and dug through the records to push back the log to 1766. As a result, Britain has the lengthiest set of rainfall data in the world.

That modest national honour has a message for those who search for patterns in the weather. Exceptional things do happen: in June 1903 it rained without stop for fifty-eight hours in Camden Square, the longest rainstorm ever recorded in England. Even so, events that are rare in the context of a lifetime are in the sweep of history commonplace. The millennium year saw severe floods, but they were more destructive in 1872. When it comes to drought, 2003 was bad but 1964 was worse. At both climatic extremes no pattern emerges, even over two and a half centuries. The half-dozen wettest years since 1766 were separated by intervals of 84, 20, 31, 57 and 40 years and the driest by 8, 66, 67, 43 and 39.

In the Pacific, too, the rains succeed or fail for no obvious reason and its inhabitants – human and otherwise – pay the price. Nowhere are the skies watched with more anxiety than upon the atolls. Many of their inhabitants have paid with their lives for the weather's fitfulness. Cocos–Keeling was ruined by a cyclone in 1909 and there was talk of abandoning the place. Tahiti suffered huge damage in the six great tempests of 1983, with a secondary battering in 1998. Now satellites search for catastrophes around the corner and provide at least a few days' notice. Many lagoon islands are sprinkled with concrete blockhouses to give their inhabitants a refuge. Their ancestors had no shelter and drowned in their thousands.

Other places are safe from storms, but face the opposite problem. The rains fail and the people suffer. In 1792 the body of the last king of Niue, an island in the south central Pacific, was found, gnawed by rats, in the parched scrub. Twenty years before, his people had been numerous and fierce enough to stop Cook from landing (the inhabitants accosted him 'with the ferocity of wild boars' and he named the spot Savage Island), but in the drought most of them had starved. On New Caledonia, with its own massive barrier reef, a dry spell half a century later led missionaries to use what they saw as biblical plagues – drought, fleas, mosquitoes, flies, famine, disease and fierce sun – to persuade the islanders towards the cult of El Niño, the Christ Child. Faith brought the promised benefits, for the weather soon improved. New Caledonia became a devout and Catholic nation, albeit one still prone to sporadic riots.

The economy of India is driven by the moods of the Pacific and Indian Oceans. The monsoon brings nine-tenths of the annual rainfall and in most years comes from June to September. The hot summer plains pull in a massive sea breeze from the distant waters. The moist air rises, cools and a storm breaks. The monsoon is an even less predictable beast than is the rain in Camden Square. From 1965 to 1987 India suffered ten droughts in the supposed wet season, but for a dozen years after that the heavens opened with no interruption. It was predicted that 2002 would be a normal year, but in fact had the poorest monsoon for three decades. 2004 was almost as bad – but just a year later the country had the heaviest rains ever recorded, with many killed in the floods.

The year 1877 marked the worst famine in recent Indian history. Six million people perished, riots broke out and the colonial power began to notice. After a second drought in 1899 the Director-General of Observatories set out to forecast when such an event might next occur. As he did, the secrets of the weather began to reveal themselves. For thirty years his men read barometers scattered through the subcontinent and across the oceans as far as Australia. The figures revealed an irregular seesaw in atmospheric pressure between the east and west coasts of the Pacific. Indian droughts were associated with low atmospheric pressure far away, while storms in Tahiti were a symptom of the same disturbance. Nobody took much notice, for in those parochial days it seemed impossible that the weather in Papeete and the Punjab could have anything to do with each other.

Now we know that this Southern Oscillation, as the shift between the two patterns of pressure is called, is a major feature of the atmosphere. It has a powerful ally beneath the waves. The entire Pacific basin is a bath-tub ten thousand kilometres wide, across which water careers back and forth in an irregular rhythm. Now and again a large slug of warm water presents itself off the tropical coasts of Chile and Peru. The warm ocean near the Americas that marks its arrival causes weak winds that reduce the amount of cold water rising to the surface. Mild and wet summers follow. At the same time, thousands of miles to the west, cooler seas prevail.

As the level of the ocean falls off the Americas it rises off Asia. At the event's peak, hurricanes hit Hawaii, Tahiti and their neighbours, while the atolls of the far western Pacific have the opposite problem, for their rains fail. As the climate shift sometimes happens near Christmas Peruvian fishermen (who feared it because of its effects on their catch) named it after the Christ Child, El Niño himself.

Those beats of the air and the ocean are nowadays referred to together as the El Niño Southern Oscillation, or ENSO. The effects go round the globe. The captain of the *Titanic*, who commented before his maiden voyage that 'I cannot conceive of any vital disaster happening to this vessel', did not know that 1912 marked a visit of the great oscillator and of a flotilla of Atlantic icebergs that set sail as he began his fatal voyage. At just that time, far to the south, Captain Scott died because the Antarctic weather was even harsher than usual.

An El Niño lasts for a year and more. Its most recent visits, in 1997 and 1998, and again in 2002 and 2003, brought floods to Brazil, heavy rains to California – and on the other side of the

Pacific forest fires made so much smoke that airports were closed. Harmful though they were, none of the Christ Child's recent calls has been a patch on the 1877 drought that killed millions of Indians.

Pacific weather records go back no further than the days of the Indian Meteorological Service and information on the movements of its waters is even less complete. Those data, limited as they are, show that the oscillation had a three- to four-year cycle from 1870 to 1910. Its rhythm slowed by a couple of years or so until the 1960s and now has a beat of four or five years. There has been little logic in the ocean's moods over that time. In 2005 the Indian government abandoned a climate model that claimed to find a pattern and set up a more complex version, which remains untested.

The record of the reefs reveals Neptune's uncertain temper over not tens or hundreds but thousands of years and more. They show that El Niño has been a most flighty child and that his older relatives were just as ill-behaved.

Coral grows faster in warm water than in cold. Its growth bands, together with shifts in the balance of salts laid down as stone production waxes and wanes, track changes in climate and show the temper of the ancient seas. In the short term, the rock confirms the efforts of the Indian Meteorological Service. A drill plunged into an atoll in the Seychelles gives witness of the rise and fall of water temperature since 1850. Every few years the Christ Child returns. His visits coincide precisely with the records of rain in India. The largest shift in the signal dates to 1877, the famine year itself. A deeper hole tracks his movements for a millennium. Sometimes, over that period, just two years passed between each visit, but sometimes El Niño failed for more than a decade.

Beneath the havoc brought by the Son of God the drills give proof of a series of more distant drummers who force their erratic attentions on the peoples of the Pacific. A longer but equally uncertain pulse led to cool periods from 1890 to 1924 and from 1977 to 1997, with a warming trend from then until today. This decadal swing has itself been an irregular caller, with no clear pattern over many centuries.

Other shifts take place on a scale of millennia. In Hawaii, corals flourish today only in the most sheltered places, although the archipelago's relic reefs show that once they did far better. The problem is the waves, for a storm can smash the work of years in a few hours. The main island's fossil reefs show that surf began to batter its exposed coasts about five thousand years ago. Before that time the changes in water temperature that mark the arrival of severe weather were considerably smaller. The shift marked the birth of El Niño himself. Until then life was calm and the polyps could grow, but within a century the peevish infant had plunged into vigorous life. Other ancient rocks show signs of an earlier ancestor 130,000 years before today. That watery deity, like his successor, took rests of several thousand years before he returned to the attack.

Hawaii is not alone in its anguish, for storm damage to coral has been endless and almost universal. The history of disaster, in climate and other ways, is better recorded in the world of the atolls than anywhere else. Their fate has forced biologists into a reluctant acceptance that what once looked like logic in life is often based on illusion. Beneath the waves and on the islands the true state of Nature is not rude Darwinian health but an endless convalescence after the last cataclysm.

After the cyclones that hit Tahiti in the 1980s parts of its outer circle of rock were reduced to rubble. In the most exposed places reefs almost never escape from that state. They can – and do – bounce back; even the scars caused by Krakatau are hard to see today. Resilient as such places are, irregular setbacks mean that no atoll can ever rest. Its life is, under a mask of order, marked by chaos. That discovery has changed the attitudes of all ecologists, on both land and sea, to their own subject.

Ecology once saw itself as a junior cousin of physics, a science in which patterns would emerge from the information gathered by mere naturalists. Its experts sorted communities into categories, just as astronomers arrange the sky into galaxies. A failure to find the right pigeon-hole was blamed on a shortage of information or – worse – on a weakness at solving puzzles. Much of the subject depended on a helpful idea called succession, the notion that all habitats go through ordained stages from a simple origin to a fully evolved end. Disturbance – cyclones, quakes or fires – might confuse matters for a while, but in time the Darwinian march of progress would continue. Now, in contrast, life in many places emerges not as a harmonious whole but instead faces constant mayhem.

People are programmed to find regularity even when chance is in control. Our outlook is so parochial and our time so brief that we see pattern where none exists. The planets follow rules but, as Darwin himself noted, the constellations, the Great Bears and Southern Crosses, are arbitrary groupings of stars invented by those who named them. In the new age of uncertainty ecologists have come to the reluctant conclusion that such randomness applies to their own science. For a time, the vague outlines of general laws may emerge but, more often than not, soon comes disillusion. Because so many variables – the numbers of plants and animals, climate, food, disease, predators and more – are involved, the danger of a strong but accidental fit in Nature is high. Reefs and rainforests often hint at brief but chance patterns that disappear when studied at more length. A central question presents itself: is life a jigsaw to be pieced together, or a lottery whose results are impossible to predict? Little by little, ecology has been persuaded of the power of chance.

That science in its new rigour now compares the abundance of plants and animals in different places with patterns generated at random by computer. What seems like logic to the biologist's hopeful eye often collapses before the cynical gaze of the statistician. Even the bizarre assumption that all species in a shallow tropical sea – polyps, fish and dolphins – are just the same in their chances of life and death generates artificial communities that look remarkably like their real equivalents, with none of the interactions with weather, food or predators long claimed to determine their fate.

Such models are not realistic, but like the notion of a perfect gas in physics, in which atoms never bang against each other, they give a standard against which to compare the real world. When tested in this way, the random view can fit the facts rather well. The power law is hard at work. Most reefs have a few common kinds of creature, more present in fair abundance and a host of species that are rare. In the same way, adjacent patches of coral may – just as expected in a universe ruled by caprice – look more like arbitrary sets of plants and animals rather than assortments of adapted types. Places a few miles apart, in conditions that appear very similar, may differ in their inhabitants to almost the same extent as a pair separated by a whole ocean.

Other ecological rules fail beneath the sea. On the Galapagos, Darwin's finches behave in the Darwinian way: species with thick and heavy bills eat hard seeds, while those with sharp and pointed beaks delve for insects. Underwater, life is less simple. The Great Barrier Reef has huge numbers of different fishes. One group, the wrasses, are as diverse as any family of birds, with hundreds of species, large and small, with massive or light jaws, teeth that grind or bite and so on. A search for pattern in their way of life reveals nothing. What they eat and how they catch it have little to do with how they are built. Even worse, the damsel-fishes, in which all species look rather the same, have almost as wide a variety of lifestyles as do the wrasses. Perhaps their home is so open to unexpected blows that pragmatism rules and they devour whatever turns up on the day.

Not everything below the surf happens by accident. A series of fossil reefs in Papua New Guinea shows that the relative abundance of some creatures has not changed for hundreds of thousands of years, proof that their fate does not depend on chance alone. The 'East Indies Triangle' around the islands of Indonesia is the richest of all landscapes for many marine forms, with three-quarters of the entire set of cnidarian species in the Pacific and Indian Oceans. Fish, lobsters and snails are also diverse, which implies that some external agent is at work on all of them. Even so, students of marine ecology now accept that many of the rules of which they were once so proud may turn on a mirage of stability in a place ruled by disarray.

The rules of that arbitrary deity mean that – as in the kitchen sink – a small shift in conditions can lead to a sudden change in state; to the collapse of an existing regime and the emergence of another. As a result, a seemingly stable ecosystem can be transformed at great speed. Men often flagellate themselves for what they have done to the undersea world but, much as we are to blame for its decline, volatility is inbuilt. On the reefs, calm is followed by a calamity as unexpected as any tsunami.

Nature's tremors, large and small, have disturbed such places throughout their history. Now they face an earthquake of human activity in which disorder has been pushed to its limits. Gradual decay has been succeeded by sudden collapse. The Caribbean has, in the five centuries since Europeans arrived, been driven to the edge. Its waters have suffered a longer and more violent attack than almost any others. Their fate shows how a healthy system can slump as resilience is pushed beyond its limits.

Christopher Columbus, on his arrival in that sea, found manatees to be as abundant as cattle at home and his successors complained (albeit with some exaggeration) that they were in danger of shipwreck on shoals of sea-turtles. The animals flourished on meadows of sea-grass that grew in the shallows and acted as nurseries for fish. At once, they were pillaged. The damage began when the turtles were killed off. The sea-grass grew unchecked and its decayed remains clouded the water and killed the polyps. The reefs coped well for hundreds of years, but by the nineteenth century the islanders had to import fish from Newfoundland.

Tourism increased the pressure. By the 1960s the main seaweed-eating fish had almost been exterminated and its job was taken over by a sea urchin. This became abundant, but then in 1983 an epidemic swept through and urchin numbers dropped a hundredfold. At once the weeds seized their chance and whole bays were engulfed by their fronds. Soon the ecosystem was dead, for the polyps were shaded out and their larvae were left with nowhere to settle. Nine-tenths of the coral of the Caribbean is now a moribund mass covered by a mat of plants.

The disaster was sudden, but the scene had been set long before. San Salvador, in the Bahamas, has recovered from innumerable storms. In 1995 Hurricane Lili smashed through. For the first time, the reef did not fight back. It had been weakened by years of abuse. The locals were shocked by the instant demise of what they had assumed to be a stable system. Ten years later the San Salvador shallows are still a weedy desert. The last straw had broken the camel's back, and when it was lifted the beast did not get up.

Chance also plays a part on a larger scale. Even before today's destruction, Atlantic reefs were less diverse than their equivalents in the Pacific, with no more than a tenth as many forms of life. A single bay in the Philippines holds more kinds of polyp than the whole of the Caribbean. Perhaps the abundance of food helps, or the vast numbers of predators, but we do not really know why that small part of the ocean is so diverse. The Caribbean's depauperate state comes from an accident of history. Three and a half million years ago the path between two oceans closed as North America clashed with its southern partner. The Atlantic found itself isolated. When an ice age struck many of its corals died and after the glaciers retreated there was no source from which it could be repopulated. One grand geological mishap set that ocean's fate and the careful workmanship of evolution could do nothing about it.

Because they are small and isolated, the most remote islands stand closest to the edge. Now and again, the power of chance pushes them over – and they may not climb back. A disease, a change of climate or a mere random shift in numbers means that life in the risky world of the furthest oceans is a fine balance between arbitrary success and casual annihilation.

A bird-watcher who follows his hobby eastwards across the Pacific has less fun the further he travels. New Guinea is filled with a vast diversity of feathered beings, some rare indeed, but as the journey goes on the islands lose their pelicans, pheasants, owls, falcons, ducks, swallows and herons. Pitcairn, five thousand miles away from his starting point, has little more than an eighth as many kinds of bird as does an equivalent area of New Guinea. The same patterns apply beneath the waves. Hawaii, in its isolation, has about forty species of reef-builder, while Palau, at the hub of the kingdom of the atolls, has ten times as many. Cocos–Keeling is as remote as any place on Earth. Its nearest neighbours are five hundred miles away in Java, or a thousand in Western Australia. Cocos has far fewer kinds of coral than does either and its waters lack the skates or rays common around most other atolls.

The power of distance is easy to understand, for a creature washed out to sea from the west coast of Australia has less of a chance of a successful trip to Cocos–Keeling than to the local reef. The ocean filters out those who cannot cross. Once land or a hospitable underwater shelter materialises the traveller may succeed in his new home, but a species that struggles on an islet will never generate a large population and will be wiped out when even a minor disaster strikes.

In the smallest and most distant places, plants and animals pass into oblivion or thrive more or less at random in the struggle against the accidents of survival and of sex. Islands act as a filter for

DNA as much as they do for species. Because populations are small they lose genes with each successive generation as, by chance, a few of those who carry them fail to reproduce. If the sole bearer of a particular variant does not pass it on the gene is gone for ever. Small groups of colonists who move from atoll to empty atoll hence tend to lose diversity on each stage of their journey until, at the end of the line, their variation is almost gone. The silver-eye is a bird with its headquarters on the Australian mainland which has reached several atolls. Birds from the islands furthest from the mainland lose much of their genetic variability.

The people of the Pacific are the most talented island-hoppers of all. Many of their homelands were not colonised until biblical times and some have collapsed again and again as their inhabitants faced disaster. The wheel of fortune turns at full speed in such places. More than a dozen tropical islands found empty by the first Europeans had been abandoned long before, some more than once. Palmerston, Henderson, Malden, Christmas, Howland, Washington, Fanning: all have robust English names but each once had a forgotten identity of its own, given by a people who did not survive.

Humans moved into Papua New Guinea some 28,000 years ago. Pottery fragments show that, around 1000 BC, they travelled from island to island across the central Pacific to Tonga and Samoa. That voyage, the distance from London to Kabul, took just two hundred years. The remote lands of Polynesia, from New Zealand to Easter Island to Hawaii, were not reached for more than another millennium. They saw their first invaders at about the time that William the Conqueror landed in Britain.

The Polynesians relive their history through their genes. The silver-eye experience – a loss of diversity through the vagaries of time and chance – was universal. By the time their journey across the Pacific ended, in Hawaii, New Zealand and Easter Island, their DNA had been forced through bottleneck after bottleneck.

Mitochondria, those ancient symbionts, pass through the egg

from mothers to daughters and have DNA of their own. They relive the history of women. Their masculine equivalents are on the Y chromosome, which descends from fathers to sons. Places filled by many people long ago – New Guinea, for example – have dozens of different types of each, both mitochondria and Y chromosomes. New Zealand, in contrast, has just four major mitochondrial groups, which makes Maori women the least variable of all humans. Needless to say, more than four females arrived when the Land of the Long White Cloud was settled eight hundred years ago but because the population stayed small for several generations some lineages were lost as their bearers had – by chance – no daughters (or no children at all).

The men of the Pacific also show the power of random loss, with just a fifth as much variability in the Y chromosomes of the remote Cook Islands as in those of New Guinea. The pattern of male movement revealed by the genes does not fit the spread of fragments of ancient pottery nor of the development of language, both of which link the Pacific's people with farmers from Taiwan. Micronesian males – those from atolls north of New Guinea – are quite unlike the males of Polynesia and neither has a clear tie with those ancient Taiwanese. The accidental loss of genes in remote places may have disguised their true path across the seas.

Simple acts of fate remove some DNA variants but, as they do so, inevitably make others more common – even if, sometimes, they are harmful. The tantrums of the Christ Child have had a large effect upon the genetic destiny of the people of Pingelap, in the storm-swept Carolines, a thousand kilometres north-east of New Guinea. There, the mishaps of history have been cruel in several ways.

The Pingelapese went through their climatic Calvary in 1775, a year shown by the coral record to be the peak of an El Niño. A typhoon, still remembered under the name of Lengkieki, struck. Just twenty survivors from a population of thousands were left on the atoll. In time the numbers built up again and tales of the disaster faded, but the genes have a memory of their own. On Pingelap an ancient fluke of climate is mirrored in the DNA of today.

In the world as a whole, about one man in twelve is colour-blind to some degree. Far fewer women are affected, for the gene most often responsible is on the X chromosome, present in double copy in women but single in men, which means that the damage is almost always hidden in females by a normal copy of the relevant DNA. The unfortunate male carriers, who lack that back-up, bear a faulty version of a certain pigment in the eye. Such people are often quite unaware of their problem: I once tested a class of students in Botswana and found one boy who insisted that a red apple and its leaves were the same colour. He was so dismayed that I have avoided the exercise ever since.

A very few children are born with a form of colour-blindness that is impossible to ignore. They see life in black and white and are so sensitive to light that they have to spend much of their time in the dark. Such people have inherited a damaged version of a gene coding for a cellular channel in the cone cells of the eye that transforms light into electrical messages to the brain. Just a single DNA base has changed. On the global scale about one person in thirty thousand has this problem.

On Pingelap one inhabitant in twenty suffers from it. The *maskun*, the 'no see', as they are called, are known to all. One hundred and fifty people are affected and as the gene (which manifests its effects only in double dose) is on a regular chromosome rather than on the X, a thousand or more must carry a single copy of the damaged DNA, masked by a normal version of the gene.

The *maskun* are common not because their altered DNA gives them some advantage, but by simple bad luck. Pedigrees show that each patient descends from one of the few men to survive Typhoon Lengkieki. That individual, the High Chief Mwahuele, was not himself colour-blind, but must have borne a single copy of the mutated cellular channel. He had many children and his descendants – with few options in their diminished community – married among themselves. The condition showed itself with the birth of Mwahuele's great-great-grandchild, the first to carry two copies of the damaged DNA.

On Pingelap a twist of fate has made a rare error common, but across most of the remote Pacific the iron laws of probability have removed variation. Just one place does not fit. The sad tale of Rapa, in the Austral Islands, a distant part of French Polynesia, contains within itself a moral about the dangers – and the joys – of life in isolation.

Distance destroys diversity, but it preserves health, for among the creatures kept at bay by the ocean are many that cause disease. Most of the pathogens and parasites that flourish in the main continents of civilisation never – until civilisation itself began to spread – made it to the furthest reaches of the Pacific.

That ocean before the Europeans was not an Elysium, but neither was it Hell on Earth. Its natives had maladies of their own. Yaws is caused by a relative of the agent of syphilis. It leads to sores and to bone damage. Human remains show that the illness was common across the Pacific (as, in a few places, it still is). Even so, scourges of the mainland tropics such as leprosy, hookworm and malaria were confined to scraps of land closest to the sources of infection. On steamy Taumako in the Solomon Islands, a few hundred miles off the mainland of Papua New Guinea, half the skeletons in a seventeenth-century burial site show evidence of such afflictions.

On Tahiti, in contrast, the natives were healthy. As Cook noted in his journal: 'Among people whose food is so simple and who in general are seldom drunk, it is scarcely necessary to say, that there are but few diseases; we saw no critical disease during our stay upon the island and but few instances of sickness, which were accidental fits of the colic.' Their remains, too, show that they were in good shape – and that, unlike Cook's own sailors, the Tahitians had excellent teeth.

Rapa – even further down the chain of settlement than Tahiti – consists of just forty square kilometres of volcanic rock. The closest land is five hundred kilometres to the north. The island was almost the last stop in the journey across the Pacific from west to east and, as in many isolated places, the genes of its mitochondria have lost much of their original variability.

Quite unexpectedly, the Y chromosomes of the men of Rapa are more, rather than less, diverse than are those of the male inhabitants of islands to the west. A closer looks shows that they come from all over the Pacific and that – unlike the female lineages – many of them descend not from Polynesians but from Europeans and Native Americans.

In the 1860s and 1870s the rains failed again and again across the western and central Pacific. Many island populations crashed as people died or migrated to Australia, to Hawaii and to the Americas. Some moved by choice but others were less fortunate. In 1862 a wave of slavery took hold and Peruvian ships – blackbirders, as they were called – descended and grabbed as many of the locals as they could. Often they were tricked on board with the promise of a religious service. So many were removed that some places were abandoned, while on others society almost collapsed.

Rapa alone managed to fight off the enemy. The crew of the slave-ship *Cora* were captured as they came ashore. Most of the slavers were sent to Tahiti for trial, but five South Americans remained on the islet's rocky shores.

After an international outcry the trade was stopped and a sympathetic, but ultimately disastrous, scheme to rescue its victims was undertaken. Four hundred and seventy of the kidnapped, who had been taken to Peru from all over the Pacific, were gathered together. They took ship on the *Barbara Gomez*, which set sail for home.

Soon, disease broke out. Just thirty-one of the refugees survived the voyage. Half were abandoned on Easter Island and the last sixteen were dumped on Rapa. As the captain said: 'He would not take them any farther: if they did not receive them he would take them back to the vessel and throw them overboard, that they might swim for their lives.'

The inhabitants of Rapa were obliged to accept the slavers' victims, but soon afterwards their own homeland suffered a terrible collapse. Its tiny fragment of rock supported two thousand souls before the arrival of the *Barbara Gomez*, but within a year just twenty native men were left (the women did rather better).

Rapa's holocaust came not from violence but from disease. The returnees brought dysentery and smallpox from South America, and the Rapanese, who had never experienced the illnesses and had no immunity to them, perished in droves. Today's Y chromosomes in all their variety tell a tale of the randomness of disaster. Most of the native male lineages were lost when their bearers died. The European and Native American chromosomes are those of the three slavers who chose to stay, while the diversity of Pacific male chromosomes is a relic of a motley crew of men snatched from their homes and abandoned far away.

In isolation the people of Rapa had found strength, for their native land, like the rest of the remote Pacific, had been safe from the waves of infection that pass through more accessible places. Once attached to the mass of humanity, the Rapanese paid a bitter price.

Epidemics are the earthquakes of the medical world. Like the convulsions beneath our feet, they follow a power law, with lots of small outbreaks, some local disasters (such as that of Rapa) and a few scourges that devastate whole continents. Just as for earthquakes, it is almost impossible to predict quite when, where and how hard the next will strike.

Nevertheless, some general rules apply. In epidemics, all men are islands and the more isolated they are, the harder an agent of disease finds it to make the journey. Any pathogen has to find a host and establish an escape route to a new home before its first target succumbs. It must also cope with the ability of the immune system to protect against a second attack.

The history of illness on islands is a perfect example of the role of chance in human affairs. To pick up a parasite is itself bad luck, but accident plays a wider role. Any agent of disease needs a pool of hosts to stay in business. If, in a small community, it kills off (or immunises) all its victims the pathogen will itself die out as not enough people remain to keep it in business. AIDS, which flickered in and out of life in small African villages for centuries, but went no further, followed this pattern. Remote places tend to be healthy. An infection arrives, kills the most susceptible, renders any survivors immune and then peters out. The science of disease, like those of genetics and of ecology, turns on the mathematics of accident.

A population in which nobody is immune and which contains a large enough pool of potential victims is at severe risk from any new illness. The South Seas after Captain Cook show the random upheavals that contagious disease can cause.

At the time of the *Endeavour*'s arrival, Tahiti was in the middle of a population boom. Captain Wallis of the *Dolphin*, the first Briton to visit (and to name the place, in Imperial style, King George's Island), wrote of his trip in the previous year that: 'This appears to be the most populous country I ever saw, the whole shore-line was lined with men, women and children all the way that we sailed along.' Within a century, numbers collapsed. On his trip to the Marquesas in 1888, Robert Louis Stevenson noted that: 'Depopulation works both ways, the doors of death being set wide open and the door of birth almost closed.' On the main island of the group – the burial ground of Gauguin – a society of six thousand people had lost two-thirds of its members by the 1870s. Disease opened the portals of death and as the Church did its best to crush local traditions it swung closed the door of birth. Cook estimated the population of Tahiti at two hundred thousand, but a French census of 1865 showed that just 7169 native inhabitants were left. Most islands did not recover their numbers until the 1960s, when Western medicine became available.

Tahiti first tested the theory of epidemics when Cook's men began to take advantage of the local sexual customs. Their pleasure came at a price. Venereal disease spread to the natives at once. It was a 'filthy distemper, which in their Language they call by a name of Nearly the same but a more extensive signification than "rottenness" in English; their hair and nails dropd off and their very flesh rotted from their bones.' Quite soon, the explorers themselves fell prey to the laws of contagion. On the journey across the Pacific, most had stayed healthy but almost all fell ill when the vessel called at the city of Batavia (now Jakarta) on the way home. Sickness was all around: 'People talk of death with as much indifference as they do in a camp; and when an acquaintance is said to be dead, the common reply is, "Well, he owed me nothing" or, "I must get my money of his executors."' Cook's vessel became a hospital ship, his crew struck down by fever and the 'bloody flux'. His noble Tahitian navigator Tupaia died and in ten weeks in that dismal port the ship lost as many men as in the whole of the previous voyage.

Other travellers faced an even worse disaster. In 1874, Cakobau, the King of the Cannibal Isles (now Fiji), ceded his lands to Britain because of the social breakdown that had followed the collapse of his society in the face of Western trade. To celebrate the event, the King and his family travelled to Sydney on HMS *Dido*. An outbreak of measles was under way, but as most of the Sydneysiders had picked up the virus while young and were immune, no more than a few died. The Fijians were less fortunate. On the way back home the King's son Ratu Timoci fell ill with the disease (he had also picked up gonorrhoea, which so embarrassed his father that he sneaked his heir past the state's own quarantine authorities).

The return of the royal party was greeted by the first conclave of all sixty-nine chiefs of the Cannibal Isles. Many caught Ratu

Timoci's measles and took the virus back to the hundreds of islands in the group, some of which had a population of just a few dozen. Within a year, a third of the hundred and fifty thousand citizens of the new nation were dead. Their villages reeked of putrefaction and sounded to the beat of funeral drums. The chiefs were furious. They suspected genocide and many of the tribes abandoned Christianity to return to the cannibal habits of earlier times.

Some European colonists saw the hand of God at work; as they said: 'fortunately these children die, but they do not die from infectious disease, they die because they are unfit to live'. However, the disaster was not planned, but was a mischance of the kind so common on small islands.

The most remote parts of the ocean faced the worst problems. In Hawaii more and more foreigners arrived and contagion came with them. Cholera killed fifteen thousand in 1803. Leprosy arrived in 1840 and was followed by influenza and dengue. There were huge outbreaks of measles. Many of the illnesses were brought by the Chinese, who were shipped in to work in the sugar plantations. In 1900 a failed effort to control a plague outbreak by burning down Honolulu's Chinatown led to the accidental destruction of large sections of the city. By the 1920s the number of indigenous Hawaiians was reduced to some twenty thousand from the million or so who once occupied the archipelago. Just one island in the chain now has as many people as lived there before Captain Cook dropped anchor.

From typhoid to tsunamis and from genes to cyclones, life in Hawaii and on its fellow islands depends in many ways on accident: on the random arrival of volcanic eruptions, storms or tidal waves, on the chance failures of reproduction and on the unexpected arrival of foreign species or alien disease. Natural selection, the agent of order, does its best to impose itself upon the chaos of Nature and society tries to do the same. Quite where the balance between the two may lie can be hard to discern. As a result, men often turn to faith to make sense of what seems a cruel and arbitrary universe.

In Cook's day, Hawaii had two opposed gods, one disposed to harmony and the other to its reverse. Lono represented reason, fertility and food, while his opponent Ku stood for turmoil, starvation and sudden war. Such discordant deities are common among nomads and fishermen, who see in life not progress but constant uncertainty. (Farmers, in contrast, most of whom live in places with more predictable climates, tend to go for a single god, albeit a celestial being prone to unexpected bursts of bad temper.)

The explorer's arrival was greeted by the Hawaiian people as the fulfilment of a prophecy: a manifestation of Lono himself, incarnated in human form as a stage in a pre-ordained cycle set by the regular rise and fall of the Pleiades. Each year, at the appropriate time, an image of the deity was carried clockwise around the island. In 1778, by chance (or perhaps, the locals thought, something more), the *Resolution* arrived on the first day of the ceremony, sailing in the correct direction around the coast. Its topmast and sail bore an uncanny resemblance to the symbol of Lono himself, a long pole with a crosspiece decorated with feathers.

Darwin has a wonderful (and politically incorrect) phrase to describe those unaware of evolution: they 'look at an organic being as a savage looks at a ship, as something utterly beyond his comprehension'. The Hawaiians were in just that position, for to them Cook's huge vessel looked superhuman. Its captain must be the Holy Man himself. The first local aboard HMS *Resolution* asked, 'Where is our Lono?', and Cook was anointed with coconut oil and fêted with pigs as the people prostrated themselves before him.

In February 1779 – just at the end of the festive period – the *Resolution* again followed the ecclesiastical timetable as it set off with great ceremony into the unknown. The natives bade farewell to their supposed god and composed themselves for the season of Ku.

Then, somewhere, a butterfly flapped its wings. A tropical storm damaged the vessel, which lost its mast and was forced to go back to Hawaii. Cook's return much discomfited the populace, who had not expected their supreme being to defy divine law in this way. Perhaps, they surmised, the Captain was not sacred at all, but a mere mortal and as subject to the reverses of fate as anybody else.

At once, quarrels broke out. Relations soon became bitter. They ended with Captain Cook's murder as he tried to recover a stolen boat: 'Captain Cook being thus mortily wounded Tumbled down the Rock into the Sea and his head fell into a gully Betwixt two Narrow Rocks. He attempted to get up could not and I believe was Suffocated in the water.' Sketches made at the time show a desperate fight, but these were later doctored to show a noble Cook, in his last moments, signalling his shipmates to stop firing on the unarmed natives.

His corpse was divided among the chiefs, who presented the Britons with a piece of their captain's thigh ('about six or eight pounds without any bone at all'), followed by a pair of salted hands and a hat. Those mortal remains were buried at sea, while the rest of his body was stripped of its flesh by the natives, who kept his skeleton.

Soon afterwards a huge outburst of steam in the crater of Kilauea killed a band of warriors as they marched nearby. Within a few months an epidemic swept through the islands. In the spirit of Dr Price, the locals identified those shocks of nature as evidence of divine displeasure. They were forced to reassess the Captain's status. Perhaps his return was not an accident of the kind suffered by mere men, but was instead part of a higher plan.

Their divinity regained, Cook's bones were kept as revered objects. They were preserved in a small basket of wickerwork, covered over with red feathers. For many years, until the old religion was at last abandoned in favour of Christianity, they led the annual Lono procession. His remnants were lost from view soon after the ceremonial visit of King Kamehama II and his wife to London in 1824 (both died soon after their arrival of measles, to which they had no immunity).

Joseph Banks, Cook's companion on his first voyage, became a strong supporter of the London Missionary Society, founded five years before Cook's death 'to spread the knowledge of Christ among heathen and other unenlightened nations'. He persuaded his Society to send their first emissaries to the Pacific. In time, the Hawaiians gave up their deities of disorder in favour of a rational and merciful saviour. Soon, their society was transformed.

Christianity now plays an important part in Hawaiian life and to many natives Cook is not a deity but the man who exposed their homeland to the evils of the outside world. In 2004 a gold-tipped cane said to have been made from the spear that killed him was sold at auction for £135,000.

The islands have, since his death, been struck many times by cyclones. Tsunamis have drowned thousands (sixty people were killed as recently as 1960, victims of the wave that followed the Chilean earthquake). Now, the ground shakes as the volcano of Mauna Loa, the largest on Earth, which has slumbered for two decades, heaves itself upwards by three centimetres a year and in 2002 Kilauea, now among the planet's most active volcanoes, increased its lava flow by thirty times. It may soon erupt again. The archipelago's inhabitants are healthier than they were a century ago – but although measles has been controlled there was an outbreak on other Pacific islands in 2003 and some carriers of the virus reached the chain itself.

An ancient homage to the random anger of the gods has been replaced by eternal vigilance. The state's medical authorities monitor shifts in the identity of measles in an effort to keep their vaccines up to date. The ocean is scanned for a temperature change that might herald a new El Niño and the Pacific Tsunami Warning Center keeps watch for the echoes of seismic tremors on distant shores. Beneath the ground, meters pick up any shift in our planet's behaviour that might presage an upheaval while, far above, satellites peer at the volcanoes to assess the risk of sudden eruption. Whether all this is enough to save the islands' plants and animals, their reefs and the Hawaiians themselves from the chaos of nature only time – and chance – will tell.

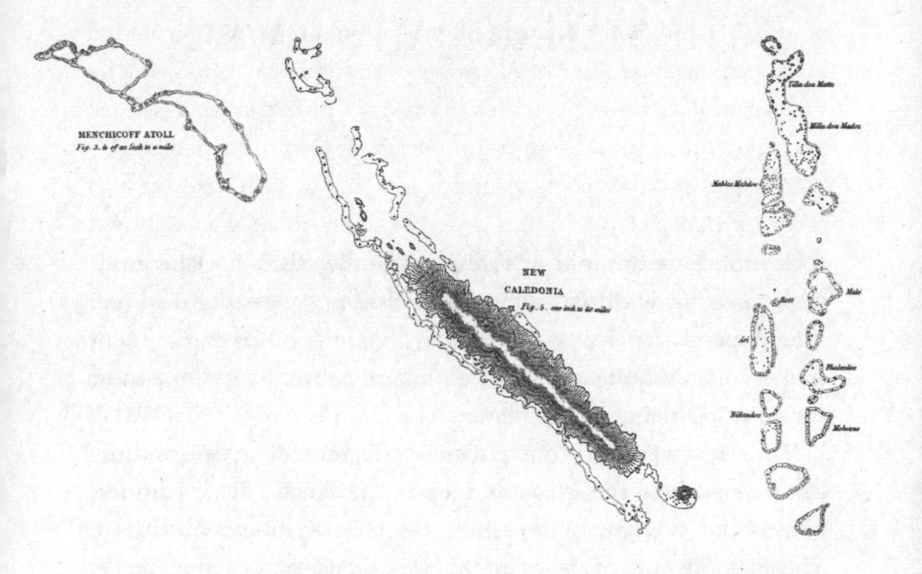

CHAPTER V

THE MAHARAJAH'S JEWELS

A diamond, we are told, is forever. The De Beers Organisation, who came up with that memorable slogan, suggests that a man should spend 10 per cent of his annual income on an engagement ring – which is quite a lot for a chunk of carbon, one of the most abundant elements on the planet.

The mantra is catchy but inaccurate. Diamonds are the product of immense heat and pressure deep in the Earth. Their hardness reflects the strength of the links between their carbon atoms. Those bonds are made as rocks are forced out of the depths through cracks in the oldest and thickest parts of its crust. In a series of global belches long ago vast blobs of the interior made a gas-powered escape to the surface. They included millions of diamonds but for almost all the future was far from forever as, in time, they were worn down to mud. No more than a few remain and in a geological instant they too will be gone.

The economics of jewellery has a lot in common with the ecology of Nature for the diamond market follows the rules that govern the passage of its raw material through land and sea. Supply, demand and globalisation apply however carbon is traded: as gems such as coral and diamond, as fuel, as flesh – or as carbon dioxide, the prime means of exchange in our planet's own chemical bourse.

Diamonds are a lesson not in permanence but its opposite. Their substance – like the billions of tons of carbon held in more prosaic form – is always on the move. The size of the element's global stock has not changed since life began, but it has faced a revolution in where it finds a home. From the earliest days its fate has been ruled by the builders of coral reefs, their ancient predecessors and their free-living marine relatives.

Carbon is a chameleon of a chemical. Soot, graphite and diamond are no more than the same stuff arranged in different ways, with atoms laid out at random or in rigid order. It has promiscuous tastes and is happy to join forces with hydrogen, oxygen, calcium and many other substances to give solids, liquids, gases and even a synthetic glass – which is how we carbon-based bipeds evolved and why limestone, a carbonate rock made beneath the sea, is among the most abundant of minerals.

A tiny proportion of the total comes as a gift from Hades to make the flashy crystals that bedeck the rich. Man's own flesh – six billion people, at ten kilograms of the stuff a head – will restore six million tons to Nature as we die. Even with the popularity of cremation (which pumps our mortal remains to where they do most harm) the dead and their ornaments play a trivial part in its economy. Carbon circulates in a great cycle, the engine of existence. Mountains and atolls, factories and forests, the ocean's depths and the air we breathe are partners in a slow pas de deux in which that dark material has a central role. Diamonds and corpses may be unimportant, but reefs ancient and modern are, and always have been, at the centre of its gyrations. Now the carbon cycle is speeding out of control.

The figures are stark. In 1958 the observatory on the summit of Mauna Loa, on Hawaii, began to collect samples of air. Four samples an hour are taken from a thirty-metre tower, far from cars, chimneys and vegetation that might produce carbon dioxide. In 1959 there were 316 parts per million of the gas in the atmosphere, but by 2008 that figure had risen by a fifth to more than 380.

Before the Industrial Revolution the skies held around 280 parts per million, which marks a huge escalation in little more than two centuries. A drill plunged for three thousand metres into the Antarctic ice-shelf tracks the gas over four hundred thousand years. Over that period – long before modern humans evolved – the levels have never matched those of today.

Different people make quite different investments in carbon dioxide. An Ethiopian generates a fifth of a ton each year, a Turk a ton and a Briton around two and a half. Each American citizen makes five tons in that time and uses as much energy as does a sperm whale (the British, too, are above porpoise level). In the context of life as a whole each citizen of the planet, however poor, is profligate, for even the humble Ethiop generates a hundred times more carbon dioxide than does a chimpanzee.

As a result of our efforts the air is likely to hold double the pre-industrial concentration of the gas by 2050, and unless something is done to prevent it that figure may rise to six hundred parts per million by the end of the present century.

That forecast is itself optimistic, for it assumes that output will not change while in fact it is on the increase, at almost 5 per cent per year. Things are about to get worse. As the oil runs out, industry will be forced to turn to the tarry sands of Canada, or to coal, whose extraction demands far more energy – and makes far more greenhouse gases – than does that of oil. Since the invention of Watt's steam engine in around 1750, coal has released 150 billion tons of carbon into the skies. On present plans the same amount will be pumped out within the next half-century from a new generation of coal-fired stations with double the power output of those at work today.

We face an unpalatable choice: change our ways or pay the price. The potential for disaster is hard to deny. If all fossil fuel reserves were burned at once, carbon's concentration in the air would increase sixfold. Even if we carry on as we are the amount will shoot up and our planet will pay the price.

Modern man has made a killing from carbon but may in the end die from it. The Sun heats the Earth but not in a simple way. The edge of the atmosphere would be a bleak place for a picnic, for a table a metre square is rescued from the cold of space by a dose of energy equivalent to that of just four hundred-watt light bulbs. The efforts of one are reflected back into the void while those of the other three travel onwards. As the planet is, more or less, in thermal balance, that too must in the end go back to space. It leaves as long-wave radiation, not visible to the eye.

If life was physics, England would have the climate of Antarctica. Fortunately, chemistry comes to our aid. The difference between a grim existence at the South Pole and the joys of the Home Counties is a result of the famous greenhouse effect, which traps radiation on its way out to space and heats up the atmosphere. The structure has several panes. Carbon dioxide is important, but water vapour is the largest element, while methane helps (and the farts of the millions of cows that feed a carnivorous world, together with the rice paddies that do the same for vegetarians, generate a lot of that carbonaceous chemical). Even elemental carbon – soot – does its bit, for the tiny particles make a black smog that soaks up reflected energy.

Hydrocarbon fuels have seen an explosion of use in the past few centuries – but carbon in other forms has also suffered great man-made swings in supply. The trade in jewels – in coral, diamonds and other precious stones – has lessons both for the natural cycle of the element and for its upheavals since the Industrial Revolution.

Once, diamonds were found only in a few river-beds in India. They were of prodigious value. The hundred-carat Koh-i-Noor – the Mountain of Light – is reputed to have been mined in biblical times and has a recorded history that dates back to 1304. For a time it belonged to the Nizam of Hyderabad, one of the world's richest men, the last in a dynasty that stretched back to 1687 when the Mughals destroyed the castle that ruled the Kingdom of

Golconda. His nation's name was a byword for wealth, for until 1725 all the world's diamonds were mined nearby. The Koh-i-Noor itself, the stone now at the centre of the British Crown Jewels, was held in the massive fortress that kept the Nizam's ancestors in power.

I once stood on the summit of that huge ruin, its walls ten kilometres around, and surveyed what Hyderabad has become: a city bigger than London, its streets jammed with traffic, cattle and lines of well-dressed children on their way to school. All tourists in India suffer a pushy guide with his dubious tales of secret tunnels and the like. Even so, a few of my escort's stories made sense.

Just inside the entrance to Golconda, beyond a gigantic portal armed with horizontal spikes to keep off the elephants, is a gatehouse with a finely modelled roof and a narrow slit used as the exit. To clap hands at a precise point in the centre gives a long series of staccato echoes that resonate through the room. An hour later, at the summit, half a mile on and several hundred feet up, a wave by my pilot to his mate in the gatehouse prompted the sound of a sudden clap, audible at just one spot, where the guard of the inner sanctum once stood. In the old days that meant trouble and a hasty call to arms. 'First telephone!' said the guide, and he was right. The echoes from below demonstrate what physicists call total internal reflection, in which waves are concentrated within a chamber and projected in a narrow beam from a single point. In diamonds, light acts in just that way. As a ray passes from crystal to air its direction changes, but as the angle becomes more acute it refracts back into the solid medium. A cut stone gathers in light from its whole exterior and after a series of internal reflections flashes fire from its face. The facets on the Koh-i-Noor match the slits in Golconda's guardhouse, but they reflect light rather than sound. The city's jewellers were famous for their skill and – perhaps – those who cut Golconda's diamonds designed the castle's alarm system on the same principle.

In 1870 the value of their product, already damaged by South

American finds, faced a terrible threat. A huge deposit was unearthed at Kimberley in South Africa. The excretions of the Kimberley Pipe had no more than one part of gemstone to fourteen million of worthless rock but such a massive effort was put in – the abandoned mine is now the planet's largest hole – that vast numbers were extracted and the price plummeted.

Faced with ruin, the rival companies merged to form De Beers Consolidated Mines. The new organisation set out to interfere with their market, with much success. De Beers' deals in carbon are small in comparison with those of Shell, Esso and their rivals, but it has managed to keep its industry under control. Its tactics are simple. The company sequesters excess production in the form of treasured heirlooms locked away at home or hidden in its own vaults. At the same time it pushes up sales with artful advertising. Its policy has useful lessons for those concerned about the increase in the amount of diamonds' raw material that blows around the heavens.

So successful was the publicity that the traditional Japanese betrothal gift of a bowl of rice wine was replaced in the years after the Second World War by an expensive ring. The real key to price maintenance, though, depended on restriction of output. With limits on the numbers of gems released, De Beers became the most successful cartel in history, with its hands on 80 per cent of the world's diamonds. Even at the time of the apartheid boycott, the South Africa-based company persuaded the Soviet Union to sell its own stones into their closed market to keep their value high. In the central bourse in Antwerp, it still fixes prices and, in spite of some twists and turns in the 1980s, has kept the business under control. From Queen Victoria's day to the new millennium, most minerals became cheaper in real terms, but precious stones got more expensive. Diamond sales have increased by forty times in dollar terms in the past half-century and De Beers' 2005 income rose to six billion dollars. The company is not as dominant as it once was – and in 2005 was forced to pay out $250 million to disgruntled Americans who accused it of interfering with free trade – but De Beers has kept its place as the largest of all diamond merchants. It still mines almost half the global total each year.

Even so, its strategy is filled with risk, and the business in crystalline carbon faces a constant danger of collapse. Already the public holds fifty times as many diamonds as are mined annually and the company's survival depends on those reserves staying in private hands. De Beers promotes its commodity's permanent worth – its value 'forever' in emotional and financial terms – to keep it out of circulation. As a result, many people see their gems as an investment. They are not, for each loses half its selling price as soon as it leaves the shop and Tiffany, the New York jeweller, has a 'strict policy against repurchasing'. Were it not for the forced rarity that surrounds that form of carbon, the rate at which it circulates would shoot up and its value would collapse.

Rarity has always been precious, and commerce itself began with baubles. The trade in coral – a compound of carbon, calcium and oxygen – flourished long before the first diamonds were discovered. The red Mediterranean form is found in noble graves that date back ten thousand years. Then, as today, the rich alone could indulge. Pliny himself complained about the negative balance of payments caused by the exchange of his local red gem for Asian precious stones and spices.

The marine stone's attraction lay in its colour. Its scarlet tinge comes from carotene, the stuff that gives flamingos their ruddy hue. The polyps that produce it live from a few metres down to two hundred or so. A mature colony makes a tree-like edifice around fifty centimetres high. Its builders shun the light and hide in caves and under overhangs in shallow water. They grow at not much more than a millimetre each year and are far less fecund than the majority of their relatives.

The submarine gem had properties that much enhanced its desirability. The Romans used powdered coral as a hangover cure, their medieval successors fed it to babies to protect against epidemics and in Italy it still works against the evil eye. The material was a universal symbol of wealth. In seventeenth-century China, the Emperor wore a red belt of Mediterranean stone, while his consort was honoured with a necklace of the same stuff. Chinese sea captains also carried coral charms, as the jewel's home in the ocean gave it the power to quell storms. In India, the lust for such finery became an obsession: 'Every oriental strives to get a string of Coral to his turban, or at least sufficient to decorate the handle of his sword. They think that to leave their dead without ornament of coral is to give them over to the hands of mighty enemies.' The trinkets of the deceased were burned with the corpse, which gave a fillip to the trade.

The history of coral and of diamonds has a moral for the carbon cycle of the modern world. It, too, has been through spectacular booms and busts and has been faced with various attempts to control the rate at which it turns. The market in industrial carbon – coal, oil and gas – has entered a new phase that could end in the collapse of the economy that gave rise to luxuries such as jewellery in the first place. Man has cashed in some long-term deposits in the carbon bank. As a result, Nature's own trade in the element is in turmoil.

Carbon has flowed across our planet since it began. It travels through land, sea and air in a great round driven by the needs of plants and animals and by the chemistry of the oceans. Coral itself locks vast quantities of the stuff into huge and hard-hearted deposits. The reef-builders' floating allies also suck billions of tons each year from the seas and skies and lay it down as rock. Long ago, their efforts brought us out of the greenhouse, and not once, but several times. They might yet – given the chance – help to rescue us from today's attempts to go back to a carbon-rich atmosphere.

The earliest indication of how connected the world of carbon must be came from the nuclear tests that confirmed Darwin's theory of atolls. At the time of the first explosions in the Nevada desert, lawyers blithely told the military not to worry about claims for damages from anybody more than two hundred miles from ground zero for it was impossible that fall-out could spread any further. Within a month the Eastman Kodak Company in upstate New York, two thousand miles away, reported that its photographic films were fogged – but why, they did not know.

Radioactive isotopes of carbon and other substances were to blame. Isotopes are different forms of an element that bear different numbers of the atomic particles called neutrons. Carbon itself comes in an abundant form known as C^{12} with an atomic mass twelve times that of hydrogen. Another isotope, heavier and far rarer, is called C^{13} and a third, C^{14}, is radioactive and decays at a steady pace back to the common C^{12}. C^{14} is generated from its lighter cousin in the upper atmosphere by the bombardment of cosmic rays and breaks down at a rate set by the laws of physics. It takes around five thousand years to lose half its radioactivity. Atom bombs get their energy as, in an instant, a heavy isotope of uranium decays into its lighter cousin. As it does, it generates powerful rays together with immense heat and unstable forms of other elements, carbon included.

Kodak's foggy film was the first hint of the enormous reach of Nature's chemical cycles. Nevada isotopes blew to Kansas, Kansas farmers grew corn and the photographic paper made from their plant's husks clouded the images formed on the products of a photographic factory in Rochester, in upstate New York. For the first time it became clear that no island – be it an atoll, a nation or a continent – is an island. Instead our planet is an entity, linked by an intricate network of winds, currents and unstable landscapes.

Just before the 1954 Bravo test of the first hydrogen bomb on Bikini Atoll, the wind changed. The blast was a thousand times more powerful than those of the first atom bombs and a cloud of fall-out descended on nearby islands. More than two hundred people were exposed. Some Japanese fishermen nearby were hit by even more of the stuff as a gritty white ash fell upon their vessel. The news caused a scandal. The Japanese government, under pressure from its people – obsessed as they were with the dangers of nuclear weapons – sent a research ship into the Pacific. There they discovered heavy isotopes not just in the water itself but dispersed throughout the food chain, more and more concentrated as the levels rose from plankton to tuna. The Japanese were the first to track the globe's carbon, on many scales. Vast quantities have, we now know, been stirred from a long slumber and are on the move. Nature operates a global economy in the element and, more and more, the industrial world does the same.

Public concern about the dangers of radiation gave politicians an unexpected interest in natural history. How many bombs, they asked their scientists, could be tolerated 'to keep the fall-out below a safe maximum for friendly populations'?

America's genial-sounding Project Sunshine claimed to be a study of the effect of calcium deficiency on babies. Its real and more sinister agenda was to use tissues taken in secret from deceased infants to count the number of 'sunshine units' – a measure of the dose of radioactive strontium taken up in the place of its natural relative, calcium. Even in Chicago, two thousand miles away from Nevada, all the children examined were contaminated. Everyone alive at the time of the tests still bears a radioactive mark of Cain in the form of carbon isotopes in their bones. After years of denial, that discovery led to a ban on aerial explosions in 1963, by which time the amount of C^{14} in the atmosphere had reached double the natural level.

The tests showed that the element and its fellows travelled far faster than forecast, through unknown pathways, to unexpected places. In a certain American dump the experts estimated that the waste would take twenty thousand years to move a centimetre. It actually migrated three kilometres in ten years. Fall-out also showed that the upper atmosphere moves particles thousands of kilometres a day in the jet stream, whose existence was revealed by the tests. At sea, too, the bombs began to reveal the movements of the ocean. In 1955 Operation Wigwam exploded a weapon almost a thousand metres under the Pacific – and within a few days a layer of radioactive water no more than a metre thick covered a hundred square kilometres. The skies, the seas and the continents were, it seemed, linked in complex and unpredictable ways.

The isotopes released fifty years ago have not gone away. The ocean itself is a vast carbon sink. It has little natural radiation (which means that the crew of a nuclear submarine are exposed to less radioactivity than is a bus driver) but, after the bombs, gained a lot more. The coral laid down in Florida, across the continent from the Nevada test site, took up large amounts of radioactive carbon, most of which is still there.

Since those days, the world's reefs have been presented with a diet richer in carbon in its various forms than at any time since man evolved. In time, they may cope, as they have so often before, but time is not on the side of those who have released the element from its bonds.

Nature's carbon currency has always been on the move, on scales from fractions of a millimetre to thousands of kilometres, and from milliseconds to billions of years. Its trade routes stretch from the ocean's depths to the summit of Everest. Most of the stocks are held in long-term accounts, some are available for instant access and a few are offered as an overdraft at a generous rate of interest. Today's pressure on the reserves may drive the whole system into bankruptcy, for as those assets are liquefied there is a surge in the elemental equivalent of the money supply.

Carbon dioxide has been around since the dawn of life and plenty still streams out of the depths. A cruel trick popular on the eighteenth-century Grand Tour was to throw a dog into a certain cave on the slopes of Vesuvius. Within seconds it collapsed, but recovered when pulled into the open. It had been overcome by gas that seeped from below (which led the ancients to see the place as an entrance to the underworld). The steamy days of the dinosaurs, the Carboniferous era, resulted from an outbreak of such infernal vents as swarms of volcanoes, excited by the clash of continents, spewed out carbon dioxide. As a result, more radiation on its way back to space was trapped by the panes of the greenhouse and the thermometer shot up.

The Industrial Revolution marked the start of a new Carboniferous. Now, space and time have been compressed and a local trade has become global. The world economy generates seven thousand billion dollars' worth of goods each year, a fifth of which are traded across borders. The cost of sea transport has dropped by three-quarters since the 1930s and that of air travel by far more. All this depends on carbon.

Our planet is a tolerant place, with plenty of reservoirs that could soak up huge amounts of the stuff. The sums involved are hard to comprehend. The ocean holds forty million million tons of carbon, the soils around two and a half thousand million tons, and land plants about six hundred and fifty thousand million tons. Vast amounts are also held in the rocks (limestone, coal and oil included), about seven thousand million tons of which are released into the heavens by industry each year.

The Earth has a long memory. If we gave up hydrocarbon fuels tomorrow (and we will not) the gaseous burden is certain to last for centuries. To sort out the balance of deposit and withdrawal has become a matter of urgency.

The North of England was once the largest commercial centre in the world. It imported cotton and sold fabrics. The first mills were powered by water and what they made travelled in ships blown by the wind. Soon the machines financed a new technology based on coal-fired boilers.

The smoke of that dismal era has cleared, but most of the gas emitted by Victorian chimneys is still floating around. Two hundred and fifty billion tons of the stuff have poured into the atmosphere since then, half of it in the past quarter-century. Two-thirds of the new carbon dioxide remains aloft (as, for that matter, does plenty of that generated by the ancient Britons who burned the northern forests).

Modern capitalism takes part in a dance of death. It is fuelled by hydrocarbons – the corpses of billions of plants, animals and bacteria that decayed in places lacking in oxygen. When brought back to life for man's convenience, their blackened remains embrace the vital gas. The union of fuel and air generates heat and, as an incidental, makes clouds of noxious fumes, together with plenty of carbon dioxide.

Oil production has risen by five times since the Second World War, while coal (a most wasteful substance) is still of central importance to the global economy, with ten kilograms burned for each American citizen each day, generating a third of the nation's carbon dioxide. Coal comes from trees and other vegetation that fell into fetid swamps but much of the planet's oil – which is made of chains of carbon up to thirty atoms long bound in loose union with hydrogen – was born at sea. Most is a product of the remains of marine bacteria and dinoflagellates. As they rot and are chemically transformed in oxygen-free conditions with the help of the sulphur found in deep and stagnant waters, they are pressed within the Earth to become, in time, black gold.

The story of oil is told in ancient reefs. The peak of El Capitan in the Guadalupe Mountains of Texas is one of the finest fossils in the world. A limestone spire soars to almost three thousand metres – a rare high point in that flat and flatulent state – with its summit supplemented by a steel pyramid donated by American Airlines. The mountain is part of an edifice built 250 million years ago. The Capitan Reef was a giant atoll that grew off the western edge of the now shattered continent of Pangaea. Most of its remains are still buried, but three large sections – the Guadalupe, Apache and Glass Mountains – have been thrust upwards by movements of the continental plates. El Capitan is a monument to a tribe of ancient builders who turned gas into solid rock. As they did they locked away a liquid fortune.

Prehistoric Texas was a chain of islands and shoals, rather like the Florida Keys shifted to the climate of Iraq. Ancient El Capitan had a few simple polyps but they were not its prime movers. The calcareous mass, like other survivors from those times, was made instead by sponges, solid algae, sea snails and a variety of other creatures able to use calcium carbonate to build skeletons. Texas chemistry helped, for now and again calcium-rich water welled out of the acid deeps into the alkaline shallows and was laid down as cement. In a rich and tepid ocean the architects worked overtime. They made a kilometre-thick bed of rock in a mere ten million years.

Then, the ocean dried and its builders suffocated beneath a layer of salt. As Pangaea broke up and the continents ground against each other, the underwater edifice was thrust upwards. In time rain, made acid by carbon dioxide, soaked in and dissolved away some of the rock (to make, as an incidental, the Carlsbad Caverns, a nearby tourist trap).

The Guadalupe Mountains are a monument both to geology and to capitalism. El Capitan's peak preserves the solid reef. To its seaward edge was a slope of rubble, smashed to pieces by the waves, whose remains have been compressed into a fractured stone, with plenty of fossils of trilobites and the like. Behind the barrier – as on many atolls today – a zone of calm and salty water laid down a rock that was transmuted into dolomite.

The Bush dynasty, with its ambassadors, governors and presidents, made its fortune by fishing for cash in an ancient ocean. The Lone Star State's reef is surrounded by sandstone and silt, while other parts of Texas are covered by a sinister black deposit. That rock is full of corpses which, as they rot, are transformed into oil and gas.

Those minerals are light and float upwards. The hydrogen sulphide that travels with them is lighter still and combines with water to make sulphuric acid, which cuts through calcium carbonate (and improved the Carlsbad Caverns as it etched its way to freedom). In many places, the gas and oil pass straight to the surface and on into the skies. Sometimes, though, the decayed remains of bacteria are imprisoned under a solid cap – a fossil reef, perhaps – and an oilfield is born.

Much of the wealth of Texas is trapped beneath the huge mass of the El Capitan formation. Two hundred thousand wells have been drilled into it and the extinct atoll still holds down fifty billion barrels of oil and trillions of cubic feet of gas. Quite how long those reserves will last, nobody knows.

Impressive as they are, the mountains of Texas represent but a modest investment in the carbon bank. Many oilfield caps are far bigger than El Capitan. Large parts of the Middle East and North Africa, now deserts, were once a calcium-rich sea, scattered with reefs and atolls. Its limestone products now sit on top of the region's vast hydrocarbon deposits. An eighth of the Middle East's reserves is held beneath a series of huge and petrified reefs. One stretches across Iraq to Iran, while another covers much of Libya.

As the oil trapped below is pumped out and burned, carbon regains its freedom. The industrial world spews out about the same amount of the element each year as is held within the limestone mass of El Capitan itself. There, tourists complain about a blue haze that comes from Mexican power stations, aided by the exhausts of their own sports–utility vehicles (and the United States generates almost half of the pollution emitted by the world's cars each year). The fumes are a small part of the problem, for the main culprit, carbon dioxide, cannot be seen.

Cars and power stations make plenty of greenhouse gases but the mountains themselves, the plants that grow upon them and the waters that cascade down their flanks have – like the natural eruptions of carbon dioxide from within the Earth – long done the same. As El Capitan's rocks are worn down by the rain, carbon, long held in a calcareous embrace, re-enters the global circulation.

The process has accelerated since the Fall of the Alamo. The forests that still clothe parts of the Guadalupes are now protected, but large sections of the Texas landscape were burned by the pioneers. They released carbon as they blazed. The dismal towers of modern Houston have also done their bit, for the cement industry makes more carbon dioxide than all of the world's airlines put together. The United States pours a ton of concrete a year for each citizen. Much of its raw material is quarried from deposits of fossil coral and roasted together with clay and other minerals. The furnaces belch out waste that helps restore the Texas climate to the hotter – if not happier – days when its foundations were laid.

There is an urgent interest in the carbon economy, old and new. How much pours from chimneys and exhausts, where does it go and how can we balance the accounts?

Ocean, land and air all play a part in its convolutions. The first is by far the most important, for the carbon cycle is intimately related to the history of the reefs. For their first billion years they lived a stagnant and acid existence in an atmosphere that held a thousand times more carbon dioxide than today. As their builders linked carbon to oxygen and calcium, they transformed huge quantities of the element into rock.

The sea is still the central bank to which all other investors turn for deposits and withdrawals. Neptune has always ruled carbon's fate. He has already absorbed half the excess poured out by man and has the capacity to take up far more. His vaults swallow up the gas in many ways. A tenth of the marine reserve is deposited in corals and in the solid parts of a variety of shell-making animals, alive and dead, while most of the rest is trapped by the chemistry of great waters.

The seas are as complicated as the continents. Their surface is carved up by slow currents and, as the bomb tests showed, their mass is divided into horizontal layers arranged by density and by age. Icy surface water near the poles pours towards the equator. It sinks as it goes until it reaches the tropics. There it heats up, rises and returns to the north or south as a shallow current.

Great streams within the depths rule many of the element's global movements. Carbon dioxide, unlike salt or sugar, is more soluble in the cold than in the warm. In addition, as pressure builds up, water can hold more of the gas (which is why champagne shoots out of an opened bottle). As a result, the ice-cold deeps swallow up vast quantities, while in the tepid shallows of the tropics the chemical and its products find it easier to escape the watery embrace and to make their way back to the atmosphere.

Neptune's intestinal movements are sluggish indeed. Most water more than two kilometres down last saw the Sun so long ago that it has never experienced the filthy skies of the modern world. As a result, half the greenhouse gas added to his domain since the Industrial Revolution is still in its top four hundred metres. It will take several centuries even for the effusions of nineteenth-century Manchester to sink into the deeps.

The marine deity takes his deepest gasps in two great aerobic centres to the east and west of Greenland and in a third lung on the other side of the globe, close to Antarctica. How much he breathes depends on the weather, for in stormy times cold water wells up and takes up more gas, while in calmer moments the carbon dioxide stays in the atmosphere. As a result, from year to year the amount soaked up by the North and South Atlantic varies by a billion tons. Those oceans are the engine of the main cold currents, which slide with their dissolved gases towards the equator. The Pacific takes up less of the element and other tropical seas may even pump it out. Many of its marine movements are still mysterious. The reefs of Cocos–Keeling have soaked up radioactive carbon from nuclear tests on Pacific atolls while others to the north and west in the Indian Ocean have not. There must be a current that moves the element to Cocos, but no further.

Wherever it comes from, a substantial amount of marine carbon goes to build skeletons. Some find a sort of immortality as ooze or rock, but most do not last for long. A rain of minute corpses falls towards the sea-floor, but two out of every three never make it to the bottom as they are dissolved in the acid waters of the middle depths. Ions of calcium and of carbonate are then released, to be used by another creature. When that in turn dies, the elements are recycled once more. Even the substance of the atolls goes back into solution and is replaced time and time again. The internal cycle of birth, death and dissolution means that most of the sea's calcium carbonate never returns to solid form but stays in solution as charged ions.

The ocean can take up a lot more greenhouse gas and over the years will absorb almost all the extra that has been emitted. The problem is time. The first reefs took a billion years to clear the air but there are not years enough in our own lives to allow their successors to rescue us from chemical disaster. It will take Neptune a century to soak up even half the excess that now blows around the globe.

The sea gains some of its raw material from the land. To cross in winter from Dover to Calais is a cheerless experience (and the return journey is even worse). Storms and slot machines are bad enough, but the Channel's brown waters add gloom to the trip. Make the journey in spring and the gales and the gamblers are still there, but the English Channel is green. It has bloomed as tiny plants feed on the nutrients that pour in from Europe's rivers. They soak up carbon dioxide. In the shallow Strait of Dover most is restored to the skies as the plants die off in winter. Further north the sea is deeper. The corpses sink into it and are swept out to join the Atlantic store.

Forests, crops and rivers have a more visible presence in carbon's affairs than do the great waters, but are far smaller players. Quite a lot of our production of greenhouse gas is hidden in the trees. In the spring, the Earth takes a breath and the amount drops as the forests bloom. In the autumn, levels rise again as their leaves rot and pour out their exhalations. Plants take up sixty billion tons or so of carbon each year, but most is stuck in a closed cycle as they exhale at night and in winter, but soak up the carbon dioxide during summer days. Tropical forests pump it out almost as fast as they suck it in, with a turnover time of five years or so. As a result they have little effect on the amount in circulation.

Soils are rich in the stuff and their stocks are renewed as plants and animals moulder away. When a virgin prairie becomes a field half its carbon is lost as its organic matter picks up oxygen and rots. It escapes upwards. To plough a pasture does almost as much damage as to cut down a forest. A survey of thousands of sites in England and Wales shows that over the past quarter-century the national earth bank has lost thirteen million tons of carbon – which is about what has been saved by all British programmes to cut down emissions in that time.

Reefs are often, with some exaggeration, referred to as the rainforests of the sea. The reefs of the land are called peat bogs. Both lay down great stores of the crucial element. The dead mosses and grasses in a bog are dense and sodden. The absence of oxygen means that their remains do not moulder away, but instead grow into huge mounds. Three hundred million years ago such material, squeezed within the Earth, was transformed into coal. Peat bogs are a huge carbon reservoir, for they hold a third of the amount in the planet's soils. They have suffered mightily. Nine-tenths of the peat of England has gone. Ireland once held a tenth of the global bogs, but its people, during their period of Celtic twilight, burned many of them in peat-fired power stations. Others were despoiled for the convenience of gardeners. The frozen bogs of Western Siberia – the size of France and Germany combined – have started to pump out methane while, nearby, the permafrost that traps carbon in the tundra's soil has begun to melt. The process feeds on itself as a pale and icy landscape turns into a series of gloomy lakes and soaks up solar energy. The decline is already so advanced that bubbles of gas stop the ground from freezing in the Siberian winter.

The trade in carbon is overheated and most people think that action is needed – but what? To keep the problem within bounds we must – like De Beers – either reduce production or find some way to remove an expensive commodity from circulation.

The United Nations agrees. The Kyoto Agreement tries to influence the carbon business from both ends. It penalises those who make it and rewards those who lock it away. In its manipulation of supply and demand, Kyoto is a mirror to the Antwerp Diamond Exchange. The amount produced is to be reduced with more efficient use of energy, while demand for the gas will be increased by mopping it up in the reservoirs that surround us.

The treaty's aspirations are modest, calling for a reduction of output of just 5 per cent in the twenty years from 1990. At the outset, enthusiasts suggested that agreements of this kind could reduce emissions by half in just a few decades. Critics insisted that our planet needs not gradual adjustment but an immediate and massive drop. In their command economy of greenhouse gas control a European tourist on holiday in Los Angeles would be hit with a carbon fine equivalent to his entire annual need for heat, light and transport. Its signatories plan an urgent rethink in 2008.

Under Kyoto, each country is given a carbon budget. Power stations, cement makers, steel companies and others are told how much gas they can release (although the scheme ignores houses, planes and cars, which pump out half the total). An excess means a fine, but a reduction gives a certificate of good behaviour. For many industrial nations, the long-term goal is below what they produce today. Britain, for example, must cut its 1990 output by an eighth by 2012. For other poorer places the threshold is set above present production. The penalties for failure are on paper severe, with a forced improvement in behaviour and stricter rules for the next round, after 2012.

The Protocol was activated in 2005, when – as agreed – it had been ratified by 55 per cent of the nations that emit 55 per cent or more of the Earth's greenhouse gases. The United States, which makes almost a third of the total, Liechtenstein (0.001 per cent) and Monaco (0.001 per cent) all insist on the sacred power of sovereignty and refuse to sign. American power stations waste heat at a rate equivalent to Japan's entire energy budget and a mileage increase of three miles per gallon for the country's cars would do away with the need for Middle Eastern oil imports altogether. Even so, the nation had, to quote its current President, 'no interest in implementing that treaty' and withdrew from the talks (although many city and state bosses have come up with local plans in an attempt to circumvent his policy). Australia, which puts out more pollution per head than anywhere else apart from the United States did not ratify the treaty until 2008.

The Russians were keen to join the De Beers cartel but less certain about Kyoto; as Putin's climate adviser said: 'anybody who is frightened about global warming is welcome to come and live in Siberia'. In 2004 they signed up. The spur was – once again – the market. The collapse of heavy industry over the previous decade reduced Russia's emissions and built up a healthy excess in its account. More successful economies to the west are hence forced to subsidise the nation with the purchase of its right to pump out unwanted gases.

Any nation that makes less than the limit can bargain with those who exceed the quota. If it costs a country more to reduce output than to pay the fine it must stump up the cash. The business makes sense, for a pound spent on improved efficiency in a decrepit coal-fired power station in Poland does more good than the same sum invested in the latest British gas-fuelled generator. The rules work for the developing world, too. If Britain installs a wind farm in Sri Lanka and thus persuades the local government not to build a power station the United Kingdom can deposit the whole investment in its own carbon bank.

Speculators can now gamble with the fate of the planet with Carbon Credits, Renewables Obligation Certificates and Climate Change Levy Exemptions. Millions of tons of the element are traded in virtual form each month – but the artificial market has not so far worked well. The price has gyrated as the permitted levels set by each nation have been marked not by long-term planning but by political expediency and self-interest. The Kyoto controls have proved far less successful than have those of De Beers, driven as the latter are by pure greed.

In the first year of the scheme most European countries set their emissions target so generously that their industries reached them without difficulty and had no need to buy credits. As a result, their value plunged. Britain set a more ambitious goal – but its polluters responded with lawsuits against the local regulator. In mid-2008, the price of a ton of carbon dioxide on the European exchange was around twenty-five euros which, given that most control schemes cost at least twice that, is not much of an incentive. The developed world's targets to 2012 are still set at absurdly modest levels despite public demands for action. The situation is confused, but most industrial countries are already well behind in their longer-term Kyoto commitments.

Whatever the treaty's success, the production of the sinister gas must somehow be limited if the oceans are to have time to rescue us from our imprudence. To sort out the supply side of the greenhouse equation should be simple: kick the hydrocarbon addiction, or find a substitute. The job will not be cheap, but with an active attack the globe could be back at today's levels by 2100, after the now inevitable high point in the next few decades. To return to where we were before the industrialists lit their boilers is impossible.

Alternative forms of energy have a part to play. The British government has thrown caution to the winds as it sets out to generate a tenth of its needs from renewable sources. The nation already has more than a thousand turbines (which together generate a hundredth of its electricity needs, about as much as a medium-sized power station) and six times as many of those angular objects are planned for Scotland alone. Britain would need tens of thousands of machines to reach its goal of 10 per cent input of renewable energy.

The French, too, are serious about their global responsibilities (and a lack of oil does make for a certain selfish interest). Forests of wind turbines have sprung up on the fossil reefs of the Massif Central. Each attracts the eye from miles away, and transforms wild country into an industrial suburb. They are admired and loathed in equal measure. *Les Verts* are in favour but the locals tend to be against because of the profits made at their expense as the electricity board is forced to buy output at exaggerated prices. The proliferation of *éoliennes* is matched by ambitious schemes for wave and tidal power.

In fact, wind farms – in France or anywhere else – have almost no impact on the problem. To insulate the roofs of five hundred houses would save more than a typical wind turbine can make. Even Denmark, long a leader in the field, has begun to abandon the devices.

France has, nevertheless, cut carbon emissions by a third over the past two decades. In the 1970s it coined the phrase *En France, on n'a pas de pétrole, mais on a des idée*s (France has no oil, it has ideas) and has lived up to it. The nation is well ahead of the game for it emits less greenhouse gas in relation to its size than any other European country – a quarter less per head than Britain (and about a third as much as the United States). The reason has nothing to do with the wind farms of the Massif Central but lies hundreds of feet below them on the coastal plain.

A red scar in the landscape half an hour's drive from my house marks the remains of the uranium mine at Le Bosc. Twenty years ago France mined more of that element than anyone else in Europe and although its own supplies have run out the nation still imports plenty. The SNCF – French Railways – boasts that its sleek trains cause no pollution, but the posters gloss over the fact that the ecological triumph is based on nuclear power. Atomic fission is responsible for four-fifths of France's electricity. The country has fifty-eight reactors and plans to build more. They make a tiny fraction of the greenhouse gases generated by a conventional power station.

Even so, nuclear power stations use a lot of energy when they are built (which reduces their real efficiency) and the problem of waste has not been solved. Finland has embarked on a new model and the British government is keen to press ahead to replace its own superannuated versions. Although public pressure elsewhere in Europe has more or less stopped the whole programme, in the rest of the world more than a hundred reactors are under construction. The price of uranium has, in response, shot up.

For nations less prepared to gamble with a radioactive future, perhaps the solution to the greenhouse crisis lies on the other side of the carbon equation – to accept that emissions are unavoidable and to try to soak up the excess.

On the rolling hills below the wind farms of the Languedoc stretches a vast and apparently immemorial forest, a grove of carbon-based life. Ancient as it may appear, the landscape is in fact new. It marks the collapse of an industry that – like the mills of nineteenth-century Manchester – failed as conditions changed. It has, since its decline, taken back large amounts of the gas that once poured into the air.

The flight from the land in France means that many of the locals now make their money from tourism. Stalls sell photographs of their villages taken from old postcards. The residents look with nostalgia at pictures of streets free of cars and unbefouled by chain stores, but the biggest change since the photographs were taken is not in the traffic but in the scenery. A century ago the rocky hills were almost bare because they were grazed by sheep and goats, or terraced into tiny vineyards. Now the land has been abandoned. In some places the amount of cover has gone up by fifteen times. After the First World War thousands of farmers moved to the cities. Soon, the trees took over and now fill almost a third of French territory – twice the coverage in the nineteenth century. The same is true all over southern Europe. As a result, the Mediterranean Forest is bigger than it has been for hundreds of years.

Forests have advanced and retreated many times over the millennia. Trees take up plenty of carbon; in France, as much as one part in fifteen of all the waste pumped out. Nowadays many have been cut down or burned, which means more pollution. In spite of the losses, satellites show that the damage is not as bad as was painted. Such places are just about in balance as ancient trees are replaced by scrub. Coppices of willow or poplar could even become a new source of fuel. Kyoto is in favour and each hectare of saplings earns a solid credit. Britain – the barest country in Europe – has a scheme to regenerate Sherwood Forest (the nation has not yet followed Deng Xiaoping's National Compulsory Tree Campaign, in which each Chinese citizen was obliged to plant three trees a year).

Pretty as they are, trees will not save the planet. A planted wood is far less of a sponge than is a natural forest with its fallen logs and tangle of buried roots. American woodlands swallow half a billion tons of carbon each year, a small fraction of what the nation produces. In Britain, even a return to the days of Robin Hood, when a squirrel could travel from the Wash to the Irish Sea without touching the ground, will not repair the damage done since the Industrial Revolution. Half of Britain's productive land would have to be used to have any real effect. The much-vaunted biofuels based on maize or sugar cane are also an irrelevance: it takes the equivalent of a hundred tons of foliage to generate the energy contained in a gallon of petrol. The grain needed to fill a car's tank with biofuel would feed a person for a year.

Excess carbon dioxide can be absorbed in other ways. Oil companies have long pumped the fatal gas into their wells to help the raw material to flow out and to restore some of their by-products to the depths. A project in Canada takes the waste from an American factory that makes hydrogen from coal, pipes it across the border and uses it to force oil out of a deep reservoir. Two billion cubic metres have already been inserted. Norway, too, pumps the material back into a natural gas field (a national tax of fifty dollars a ton on released carbon explains why its oil industry is so keen on ecology). To store even a tenth of the pollution a volume of gas equivalent to that of the world's entire trade in oil would have to be forced into the ground each day.

In Holland, a refinery donates its waste to hundreds of glasshouses used to cultivate roses, where the extra carbon dioxide speeds their growth. Vast amounts could, in principle, also be stored in soil: the hugely fertile patches of loam found in the Amazon basin are the relic of a dense human population that dumped smouldering rubbish. Over the years the earth accumulated twenty times more carbon than that present in unfertilised soils nearby – more, indeed, than in an equivalent area of forest. 'Biochar' based on farm waste or paper could, its enthusiasts claim, trap almost as much of the element as is emitted by the world's power stations, as well as restoring our impoverished soils to a more natural state.

The most ambitious schemes to calm the carbon cycle depend on the biggest sink of all, the sea. Enthusiasts plan to improve upon the efforts of the coral-builders and turn air into solid rock. Can man put his wastes back into Neptune's geological store, where they belong?

One idea is simple: to pump the gas or to freeze it into dry ice and deliver it to the high-pressure depths. Five kilometres down, deep below the sea bed, the chemists promise that the pressure will trap carbon dioxide molecules in minute cages made by molecules of water. They will be unable to escape. In principle – a long way from practice – a patch of ocean floor no bigger than Inner London would be enough to soak up the annual output. Other schemes use chemistry to mimic Nature. They take carbon dioxide from factories and mix it with elements such as calcium or silicon to form inert minerals that can be dumped as synthetic reefs. The science is simple and the raw materials easy to find. The job is, alas, far too expensive to be feasible.

A more subtle idea takes advantage of an unexpected shortage of a certain raw material in the seas. Darwin noted that the azure waves around his atolls sparkled in a manner quite unlike the turbid waters of the British Isles. The distant Pacific is blue not from cold but from starvation. Very little grows because its waters are sterile. The ocean is not short of most of the ingredients of life, for nitrogen, phosphorus and potassium are present in abundance. The missing element is iron, which is needed in minute quantities to keep plants alive. Even the Atlantic suffers from iron deficiency at the height of the spring bloom.

That mineral can reach the remote oceans only when blown from the land. More is available in the dry and dusty climate of an ice age and, over the past four glacial cycles, wind-borne iron has led to an increase in the number of green plants in the sea, more calcareous skeletons under the surface and – as a result – less carbon dioxide in the skies.

Twenty years ago came an audacious plan, to seed the seas with the absent element in the hope that they would bloom. As the new crop of tiny plants and the animals that feed on them died, their carbon would sink to the bottom, be transformed into rock and the climate crisis would be over. 'Give me a tanker of iron,' said the man who came up with the idea, 'and I'll give you an Ice Age!'

The scheme sounded Utopian, but a rich environmentalist tried it from a yacht and it worked. Since then, the experiment has been repeated several times, on larger scales and in different places. Almost always, a dose of iron causes a green explosion in the sea.

The numbers do not, alas, add up. The break-even point is ten dollars for each ton of carbon removed, but the actual price is ten times that. Chemistry suggested that each ton of metal would sink a million times as much of its fellow element, but that was far too hopeful. In the largest experiment, in the Antarctic, the iron generated no more than a few thousand times its own mass of living carbon. The seas bloomed, but for just a few days. Even worse, just a small part of the rain of corpses sank far enough into the depths to allow their substance to join the permanent store. The rest dissolved and the carbon leaked back into the atmosphere. On those figures, armadas of ships full of iron would be needed to soak up our output of the fatal gas – and they would have to sail to a mythical ocean bigger than the Pacific.

All these methods are ingenious, expensive and more or less hopeless. The carbon crisis might instead surrender to the simple rules of capitalism, for as oil, gas and coal run out their price will rise and the system will limit itself.

The market has already done a lot. Modern gas-fired power stations are more efficient than their ancestors and cars are even better, while new houses use half the energy of those of fifty years ago. However, as houses, cars and planes improve we use bigger cars, buy more air-conditioners and take longer holidays to match. As yet the system shows little sign of limiting itself, with the real price of hydrocarbon fuels still below its 1980s peak. Nobody knows when the oil will run out. The United States Department of Energy estimates that enough is left for thirty years. Others put the reserves at a third of that level and suggest that the time to worry is now.

Prophets of doom point out that Shell Oil has been forced to reduce its estimates of reserve size, while those of delay point to the many unfounded claims that ecological disaster is around the corner – and insist that, in any case we are a tropical species who moved to the cold not long ago. Many of us spend half our income in the endeavour to stay at a comfortable temperature, and could cope with ease with the climate of the dinosaurs. Capitalism, they say, will provide its own solution.

Carbon, in the form of coral and diamond, has already tested the strengths of that potent mechanism. For one of those commodities an overheated economy was restored to a calmer state, but for the other irrational exuberance still prevails. The fate of the planet will depend on whether the hydrocarbon fuels follow coral, or diamond, into the future.

Nowadays, most coral jewellery is cheap. The raw material comes from all over the tropics and the finished product is made in local sweatshops. Neither the fishermen nor the artisans make much money and even the merchants are no more than moderately rich. Its fate has a wider lesson, for just before the Industrial Revolution a tremendous boom in the red trinkets was followed by a collapse as the financial rules came into play.

Four centuries ago came the Coral Rush. Huge fortunes were made as Mediterranean gems were shipped to India and China. Demand soon outstripped what the fishermen could harvest; as a 1644 report from Goa said, 'So great a quantity cannot be produced in Europe yearly as would vend here.' The jewels soared in value.

At first, the sea gave up vast quantities and in the seventeenth century it was 'common for the coral fisheries to bring from the islands adjacent to Leghorn from six to eight thousand pounds of coral each boat. About three hundred such boats are employed.' The harvest was dangerous and wasteful work. Planks formed into a heavy cross were dragged along the bottom. On a good day, fifty men and a dozen boats were needed to lift the mass to the surface. Much of the catch was smashed. Soon, the raw material became harder to find. It was then collected by divers, many of whom drowned. The finest beds were near Barcelona. In the 1850s a young Catalan invented the first true submarine as a technical aid to the industry. Narcís Monturiol's 'Icti-Neo' looked like a fish, with motors in its tail and fins to control direction. Its hull was powered by oarsmen and the vessel used quicklime to soak up their exhalations. Investors were told that they could recoup their cash in a week from virgin beds in the deeps, but Monturiol soon went bust.

So insistent was the call for the precious stone that an ingenious trade began, a mirror of the natural cycle, in which one form of carbon was exchanged for another. It involved the barter of European coral for Indian diamonds.

The business was under the control of the Jewish community. Many Portuguese Jews who had fled the Inquisition became involved in the London Diamond Exchange. A few moved on, to the Portuguese colony of Goa and from there to other places in India, Fort George – later known as Madras – most of all. Its English governor, Elihu Yale, married a Jewish woman and gave a large donation to the Connecticut College in New Haven, now a university more often referred to by another name. A later incumbent, Thomas 'Diamond' Pitt, grandfather of the first great statesman of that name, obtained by devious means the largest stone ever seen, which in time found its way into the French crown jewels.

The diamond merchants soon became interested in the coral trade. Both commodities took advantage of the irrational desire for self-aggrandisement and both were expensive. Even better, the red rocks were bulky and went out to the subcontinent in ships otherwise empty, while packages of diamonds could be hidden in the vessels that came back to Europe packed with spices, carpets and the rest: 'The coral exporters give security, if required, that the produce be brought home in diamonds, diamond boart, musk, ambergreece or bezoar and in no other goods whatsoever.' In the quarter-century after 1750 licences were issued for the import of marine jewellery to India valued at over a million pounds and the export of diamonds (and India was then the sole source) worth even more.

For many years both ends of that global cycle were profitable, but in the late eighteenth century the coral business began to wither. New supplies became available from tropical seas while Indian fashions changed. Within a few years the trade in decorative marine stones collapsed.

Diamonds, too, ran into problems. First came new mines in South America. The various wars of the early nineteenth century led to a severe decrease in demand. As a Parisian jeweller put it: 'In order to sell diamonds, people must be at their ease, money must circulate and trade must be settled.' The price began to sag and it seemed that the gems would suffer the fate of their submarine relatives.

For diamonds, De Beers came to the rescue and manipulated the rules with much success. It will not be able to do so for much longer. Science has faced their business with the problem that killed off its maritime twin: an excess of raw material. Market forces are adamant indeed and are tied to a hard word, for adamant, the Greek for 'invincible', was, in Middle English, used for diamond itself. The gem's rigidity and its glitter are related, for both come from regular and firm bonds between its atoms. From Plato until today the stone lived up to its name: it could scratch anything, but nothing could scratch it.

In 2004, diamond was vanquished – by diamond. Hydrogen and methane at two thousand degrees Celsius and seventy thousand atmospheres of pressure were blasted in a powerful microwave. A tiny seed of real diamond was included and at once a rain of carbon atoms began to arrange itself upon that framework. When the sample was tested on a diamond anvil the anvil shattered, for the new material was half as hard again as its natural rival. It is a more perfect and regular form of the crystal. For diamond, a brand new source of a product better – and far cheaper – than the original has become available.

That, needless to say, does not cheer those whose business depends on natural stones. Previous efforts to sell artificial diamonds – some of which are quite realistic – were attacked by De Beers' Gem Defensive Program, which used ultraviolet light to peer into their structure and to reject those made by man. The new microwave versions pass the test with no difficulty and the traders have been reduced to the somewhat desperate claim that 'If people really love each other, they will give each other a real stone.'

Soft hearts make for hard cases and romantics are unlikely to fall for that ploy. The new versions are available for a hundredth of the price of their rivals. Soon after they became available an Indian trader bought a thirty-five-thousand-dollar batch, which he sold in his native land at a large profit (and with no needless fuss about where they came from).

For diamond, as for coral, the supply side is set to triumph. De Beers may as a result share the fate of an earlier monopolist in the carbon trade, the huge cartel that for a time controlled both the coral and the diamond business, to its immense profit. The East India Company was founded in 1600 as a tool to rule the sub-continent. Cecil Rhodes, himself a founder of the De Beers Mining Company, described his organisation as an African version of the East Indiamen and planned, like them, to use its commercial power to fund a series of political takeovers. The Mughal emperor Jahangir had asked James I to 'command your merchants to bring in their ships all sorts of rarities and rich goods fit for my palace' and the British were happy to oblige. From that business in fripperies – with jewellery a major item – the Company grew to become what Macaulay called 'the strangest of all governments, designed for the strangest of all empires'. In 1698 the British set up a rival agency to introduce competition, but, in an early version of the De Beers philosophy, the monopolists foiled that by buying its shares.

In 1850 the East India Company was at the peak of its powers. To mark its two hundred and fiftieth anniversary it presented that enormous stone, the Koh-i-Noor, to Queen Victoria. Eight years later, to its own amazement, the Company collapsed, its stranglehold over trade in gems, gold and other goods undermined by the new power of industrial capitalism. Nowadays it owns a small office in London and holds the exclusive right to the sale of tea on the island of St Helena.

After the departure of the British in 1948 India embraced a form of Hindu twilight in which the spirit of monopoly ruled once more. In the tradition of the East Indiamen, the economy was managed, outsiders were kept at bay and investment was under central control. Tariffs deterred foreign commerce, growth was at a snail's pace and the country stayed poor.

In the 1980s the subcontinent began to embrace free trade. Business boomed and now grows at around 8 per cent a year. In legend, the streets of Golconda were paved with gems. Those days have returned. Hyderabad, the conurbation which surrounds that ancient citadel, has a prosperous middle class and is a global leader in computer science, biotechnology and telecommunications. Its own seams may be exhausted, but the Indian diamond industry now processes 90 per cent of the world's stones in what has become the country's second-largest export earner. Hyderabad itself plans a new souk to sell baubles to its bourgeoisie. China, too, has galloped down the capitalist road. In terms of investment it comes second to the United States, even if its vast population means that most people are still poor. It expects to quadruple its national wealth in the next fifteen years.

All this needs hydrocarbons. Since 1980 both nations – the planet's largest and second-largest – have tripled their use. India has ten times as many cars than in those days and the new middle class uses fifteen times more energy per head than do the poor. China has followed the same path. Every two years it builds power plants with an output equivalent to the entire British national grid, which means a large new coal-fired station a week for the next five years. It already uses more of that filthy fuel than Europe, Japan and the United States put together. The dust and grit now reach California, and China's output of carbon from coal now exceeds that of any other industrial country.

The nation has made some effort to hold back emissions – Beijing has the globe's largest fleet of gas-powered buses – but is growing so fast that the amounts are nonetheless certain to soar. It now comes second in the world as a generator of carbon dioxide, with India not far behind – and India's population, if not its industrial might, is set to outstrip that of China by 2030.

In the 1992 Kyoto Agreement each was classified as a 'developing nation', free from any obligation to reduce output. That infuriated the Americans. Why should they shackle their industries when their rivals are allowed to get away with the carbon-based equivalent of murder? The excess produced by China alone in the next quarter-century will surpass the Kyoto target five times over. This is, they insist, not free trade but an attempt to rig the economic rules.

The Americans used the exemptions as an anti-Kyoto alibi. As a result, the industrial carbon cycle is in full flow, and greenhouse gas is rising even faster than before. Forty per cent of global energy still comes from oil, a quarter from coal and the same amount from natural gas. The amount used is set to double in the next two decades.

As the excess pours into the skies we face an interesting dilemma: whether to marry the demand for fuel with a safe level of supply, or to use up our reserves and pay the price. If we do, man's memorial may well be laid down – as in the coral mountains of Texas – in the element itself: as a thin black layer in the geological record that marks the passing of a short-lived species that tried, and failed, to defy the rules of Nature's economy.

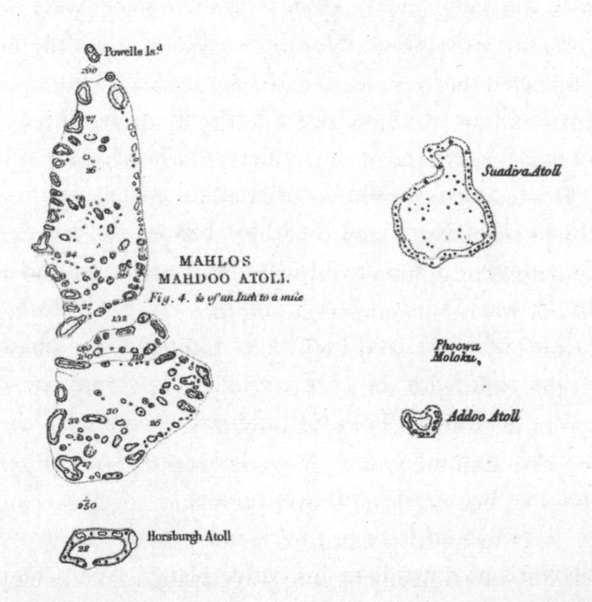

ENVOI

A PESSIMIST IN PARADISE

The Garden of Eden was an oasis in a desert, but in the 1930s the film industry turned it into an atoll. The first underwater documentary, made in that decade, was shot on the Great Barrier Reef. It caused a sensation even after the British censor cut out pictures of turtles laying eggs on the grounds of indelicacy. More and more tourists visited the 'Lucent gardens of Neptune, which rival those of Hesperus, the daughter of Oceanus' as such places were – with a certain restraint – described. Numbers exploded with the invention of the aqualung in 1942. Jacques Cousteau, who came up with that marvellous machine, was the first in a long line of television impresarios to exploit the underwater world as a subject. Now its lucent gardens are almost as familiar as our own.

Since those days, Neptune's estate has become an icon not of Nature's beauties but of her decline. Its fate is a stark reminder that the end of the world in its natural condition could be closer than we think. Two millennia and more ago, Horace came up with a motto to help cope with the certainty of decay: as he says in his prescient way, *carpe diem*: 'Envious time has passed: seize the day and put no trust in the morrow.' Many coral reefs have rather few days left to seize before they are lost for ever.

Today's gloom about their prospects marks the latest in a series of reversals in man's attitude to his native planet. An ancient fear of the wild matured into the Enlightenment's desire to explore it and then to the romantic notion of a universe in harmony with itself. Such views have been replaced by an obsession with a vanished Arcadia and with the fear of some coming plague that will mark the end of the reefs, the rainforests and – perhaps – of ourselves.

Mankind is commoner, dirtier and greedier than ever before. He rules over his domain with little thought for the future. Tropical shores are polluted, pillaged and ravaged by disease. They are mined for stone, dynamited by fishermen, poisoned for the aquarium trade and suffocated by filth. The atolls are in real peril as the world heats up and the waters rise. They face, in addition, a great triumph of the average; a global homogenisation as Nature retreats in the face of human expansion. *Homo sapiens* has reversed half a billion years of geology. He has reunited Gondwanaland, not by closing seas but by crossing them.

My first visit to a coral garden, in Hawaii, was as a result somewhat of a disappointment. The state's motto is 'The life of the land is perpetuated in righteousness', but few of its plants and animals live up to that confident maxim. Hawaii is a paradise filled with pessimists (who can, if they wish, patronise its two hundred businesses called Paradise, one Found, but none of them Lost). The island chain has become a microcosm of the ancient balance between romantic fantasy and awkward fact.

On his third and fatal trip Captain Cook reached the archipelago and at first much admired it. Within a few weeks he became plagued by cynicism about the natives, guilt about the evils wrought by Europeans and hatred for his shipmates ('Curse the scientists and all science into the bargain!'). His journal bemoaned his crew (whom he flogged with new enthusiasm), the locals (cropped ears were a useful punishment) and even the point of his journey. His mariners remarked the change: 'I can't well account for Capt Cooks proceedings on this occasion as they were so very different from his conduct in like cases in his former voyage.' Cook owed his plunge into despair as much to a gut blocked by round-worms and to the heavy use of opiates as to the loss of a flawed Elysium, but many of his heirs have had their illusions shattered by the truths of the Pacific islands.

Within a century of the first European visitors, Hawaii's decline was well under way. In 1866 Mark Twain saw the chain as 'the loveliest fleet of islands that lies anchored in any ocean'. Its way of life was already about to collapse: 'The traders brought labor and fancy diseases – in other words long, infallible destruction, and the missionaries brought the means of grace and got them ready.'

Hawaii's culture may have been in decline but at the time of Twain's visit many of its plants and animals were still extant. A lot has changed since then. To walk through the forests today is to feel oddly at home, and not just because of the Union Jack in the state flag, a remnant of the time before sugar barons annexed the kingdom to the United States. On land and sea it is possible to see more global diversity in an afternoon than any ecologist should expect. Parts of the scenery look like a student hall of residence gone mad, with house plants such as the Swiss cheese plant and the weeping fig present in vast exuberance. English sparrows, California quail, Indian mynahs and Australian cockatoos abound. Even the exotic red-crested bird that once starred as an endangered native in the movie about the dangers of alien imports shown in the aircraft before touchdown is a native of Brazil.

The Hawaiian chain now holds around twenty-five thousand species of plants and animals, more than a third found nowhere else – which gives the islands far more unique species than the Galapagos can boast. They still have stinkless stink-bugs, flightless flies, carnivorous caterpillars and no-eyed big-eyed spiders. In Mark Twain's day there were many more such creatures. The archipelago has, thanks to man's attentions, suffered more extinctions per square kilometre than anywhere else on Earth.

One group close to my own scientific heart has paid a heavy price for human progress. Their loss has ruined the chance to test a tenet of evolutionary theory. In 1905 the Reverend J. T. Gulick published a study of the islands' unique snails and pointed out for the first time the role of chance in evolution. How, he asked, could ravines a few yards apart each evolve a new species by natural selection alone? Accident must surely be important as, at random, a certain variety became common until at last a new form of life emerged.

That idea informs much of the modern theory of evolution and it would be marvellous to test it on the good Reverend's originals. Alas that is impossible, for almost all his subjects have been lost. A few hopeful biologists – myself included – still hunt now and again for the remnants that cling to the basalt peaks. I helped to sample places on the Big Island of Hawaii last visited by collectors a century ago. Most were filled with alien forms, many from Europe. In a remote spot on the slopes of the Mauna Kea volcano, in a native forest of scrubby trees (unspectacular, but far more precious than the imported vegetation popular with visitors), emerged a single live example of that small, grey, noble and underrated mollusc *Succinea konaensis*, among the rarest of all snails. Its DNA puts it in the Hawaiian family tree and says when and where it evolved, but like its fellows on islands around the globe the animal will soon be gone.

The same is true beneath the Hawaiian seas. Almost four hundred marine species have come in. Even the State Gem – the black coral, which once supported a thirty-million-dollar trade – has been smothered by an Atlantic relative that can grow a centimetre a week. In some places nine-tenths of the indigenous coral was destroyed within three years of its appearance. On the shore, seaweeds introduced to help the agar industry form large piles, choke the natives and must be removed at great expense.

The state's reefs face many other stresses. The US Navy blasted a huge channel through them to open up Pearl Harbor. Others were destroyed for the trinket trade. Nineteen out of twenty American visitors dive (the Japanese are less anxious to get their hair wet) and, as they do, stand on some of the finest structures and ruin them. Visitors do plenty of indirect damage too, for the sewage outlets built to treat the effluent of Honolulu poison the local seas. On other islands pineapple farms pour in silt that kills off marine life.

Above and below the waves Hawaii and other elements of the coral empire are showing real signs of strain. A great disaster is under way. The figures are far from reliable, but one in ten of the world's reefs is already dead or almost so, a third remains in critical condition and about the same proportion is under threat. Just one part in four remains more or less unscathed. Real prophets of doom see a loss of more than three-quarters of all such habitats within the next century.

Today's catastrophe has happened before and without human help. The end-Permian extinction of 251 million years ago, the biggest the Earth has ever seen, has an uncomfortable resonance with modern times. The crisis took hold with dreadful speed after an era of stability that had lasted for millions of years. In less than a hundred centuries – an age in history, but a geological moment – it changed the oceans for ever.

The calamity was caused by a silent but now familiar enemy. It killed 90 per cent of all marine animals and did even more damage on land. A carbon crisis was to blame. As the continents drifted and clashed a mass of lava was forced from cracks in the surface. Its visible remnants are the 'Siberian Traps' (*trappe*, the Swedish for staircase, refers to their stepped structure). An area bigger than Europe was covered in black basalt more than a kilometre deep. As the ground heaved it poured out carbon dioxide, which was accompanied by a huge belch of methane as the permafrost melted. Over a few centuries the temperature shot up by six degrees and the planet almost died.

As the gases built up carbon reached levels five times higher than today (and even the gloomiest modern forecasts do not predict that). Forests and green plants perished and the coal production factory closed down. The collapse fed upon itself. On land, a lush terrain became a desert. A few fungi and simple plants, the ancestors of today's club mosses, were the sole survivors. Fossils of their sex cells show that quite often their development went wrong. Mutations caused by an increase in ultraviolet as the ozone layer was destroyed by the toxic gases emitted by volcanoes may have been to blame.

The sea suffered as billions of tons of soil poured in. The evidence is preserved in the form of fossilised marine silt filled with scraps of wood and the remains of green plants. Before the crisis these were broken down on land by bacteria, but after the great deforestation – which left a landscape of rubble as evidence of the floods that followed – their native soil was washed way. The dead vegetation stifled the oceans, which turned acid, stagnated and stank. They became almost sterile, and as the globe warmed the currents slowed down, the ice melted and the waters began to rise. Soon most reefs were drowned.

A drill through an extinct Chinese reef of those days shows the extent of the disaster. Before the crash it was filled with hundreds of kinds of polyps, sea-shells, algae and trilobites. Within a geological instant they were gone, leaving just a few worms grubbing for food. The production of calcium carbonate fell to a hundredth of its previous level and the great limestone edifice, together with many others (the Capitan system of Texas included) perished almost at once. Several marine plant and animal groups that flourished before the cataclysm were lost for ever, and cnidarians themselves have never regained their earlier dominance. A few of the stone-making survivors reverted to a simpler existence as anemones and have stayed that way ever since. For five million years after the end of the Permian, marine rock was made by microbes, as it had been in the first days of life. In time the solid polyps bounced back, but it took them far longer to recover than from the explosion that killed the dinosaurs sixty-five million years ago. After that nasty shock many corals returned to full production within a hundred thousand years.

We are not yet in some new Permian, but signals of a similar calamity are all around. Each of its ancient elements has a parallel today. Levels of carbon dioxide are set to double and the temperature to increase by three degrees or more in the next century or so. As it does, ice sheets will melt and the sea will rise. Already many shallows have filled with silt as the forests are destroyed. The ocean, too, shows signs of disquiet, for the deep currents have begun to shift and its waters are warmer and more acid than before. Wherever they are, desert islands and those who live upon and around them have begun to notice. They have good reason to worry.

Today's problems are not due to the movement of continents but of people. The economic boom has as much potential to kill off the natural world as did the geological upheavals of two hundred million years ago. Since Captain Cook, human numbers have increased tenfold, from six hundred million to over six billion. It took from the origin of the universe to the Wall Street Crash to put the first two thousand million souls on Earth, from 1929 to Richard Nixon's resignation in 1974 to repeat the job, while the next couple of billion arrived in time for the Millennium Dome. As everyone over sixty knows, the high point of sexual history was in the 1960s – the peak of reproductive success. Those happy times had an annual growth rate of more than 2 per cent.

Today we multiply at half that speed and most countries have a fertility level below what is required to maintain their populations. Europe is at the bottom of the reproductive stakes but has plenty of fellows in restraint. They include China, Brazil and large parts of India. In just a fifth of the world's nations – in Pakistan, the Arabian Peninsula and whole swathes of Africa – do families still average four or more. *Homo sapiens* has, since the invention of agriculture, become ten thousand times more abundant than any non-domesticated mammal of comparable size. We may glory in our profusion, but soon we will set off on the long road back to rarity, for by 2060 human numbers will peak at nine thousand million.

Our decline will be too late for the polyps. Huge damage has already been done and matters are certain to get worse, for the coming increase in numbers will not be evenly distributed. Many northern countries are set to shrink in the next half-century (Russia by a third) while in the tropics the population will go up by four times.

An animal that evolved in the land-locked plains of Africa has moved to the seaside. A quarter of the globe's inhabitants – more than a billion people – live within a hundred kilometres of the coast and that proportion will double within two decades. Almost half of all new houses in the United States are built near the ocean. In south-east Asia – the capital of coral – more than half the populace now lives within twenty kilometres of the shore. All this spells real danger for the underwater world.

In the Permian the reefs were the first ecosystem to warn of a more general catastrophe to come. Now they face an enemy more ruthless and far more effective than those of 250 million years ago. It takes as little heed of the consequences of its actions as did its primitive and mindless ancestors.

The sea, like the air, is a common good, but billions of people have never seen it. Until not long ago the oceans and their margins seemed secure against the worst that man could do. People treated them as hunter-gatherers did the land, as a world without fences, a tract open to all. Science supported that optimistic policy. With fine Victorian confidence Thomas Henry Huxley – the man who first realised that polyps, jellyfish and *Hydra* are members of the same group – claimed that: 'Any tendency to over-fishing will meet with its natural check in the diminution of supply . . . We advise that all Acts of Parliament which profess to regulate, or restrict, the modes of fishing pursued in the open sea be repealed; and that unrestricted freedom of fishing be permitted hereafter.'

Huxley's logic remains popular. I myself witnessed an act of ecological pillage that paid homage to his buoyant philosophy. I set up a research project on the wind-battered island of San Miguel, off the coast of California (the experiment involved the release of millions of mutant fruit flies, but became an expensive fiasco). The Channel Islands, of which San Miguel – now uninhabited – is one, had been the home of a community of American Indians, the Chumash. Their livelihood lay under the sea. Mounds of abalone shells, millions strong and thousands of years old, still lie behind its beaches as proof that the ocean can, given the chance, feed a nation.

Thirty years ago the tough but tasty molluscs became fashionable among the rich and the Great Abalone Rush began. The Indians had hunted with poles, but aqualungs made it possible to plunge deeper in search of the large flat shells of the prized white abalone. In 1970s California the rule then, as now, was scratch a hippy, find a capitalist. Most of the bearded youths in the wretched ghetto of Isla Vista, near Santa Barbara, were out to make a quick buck. Some grew marijuana (those were the days before twenty-year sentences for possession), some went in for sexual healing and a few turned to the seas.

Abalone were abundant on rocky reefs from Point Conception, north of Santa Barbara, to Isla Asuncíon in Baja California, six hundred miles to the south. Many places had one snail for every square metre – ten thousand a hectare, hundreds of millions altogether. The first divers made a fortune.

A dead shellfish soon rots so it was essential to get the catch to market as soon as possible. The fishermen used powerful speedboats with almost no shelter from the weather. I often hitched a ride to the islands. On stormy days it was hair-raising, as the prime rule of navigation was to steer around the breakers at twenty knots. Many times I swore, as I held on to a bouncing boat in an icy spray, that I would never do it again, but I gave up only after a violent political row with my crew about whether anybody, apart from the hippies of Isla Vista, had ever come up with the idea of putting joint money into a pot to cover the costs of health care.

Matters soon became tense between the fishermen themselves. Good sites were jealously guarded and members of the peace and love brigade began to take pot-shots at their rivals. By 1978 so few abalone were left the fishery collapsed. A quarter of a century later the stock has still not recovered. Around the islands the animals are ten thousand times less frequent than before. Just a few hundred are left. Like the polyps, abalones spawn millions of eggs at a time, but they have not been able to re-establish themselves and a planned breeding programme failed when only one female was found. Now so few remain that most eggs never meet a sperm and the white abalone, which once fed California, was protected under the Endangered Species Act in 2001. Its extinction cannot be long delayed but the delicacy itself is back on the menu at ten times the previous price, flown in from New Zealand, where another abalone awaits its end.

In many places corals have suffered the fate of the white abalone, and for the same reason. Their vulnerability became obvious as soon as the red Mediterranean version was traded to India. After three centuries of exploitation just a few colonies of that elegant animal are left and most are so small as to be not worth harvesting. After a burst of new stock from new beds in Algeria and Morocco in the 1980s the harvest has dropped by two-thirds. Now almost no raw material is available and it may take a hundred years for even the protected populations to return to commercial size.

The trials of the atolls began with the first travellers across the Pacific. People got to most of Polynesia two or three thousand years ago. The settlers depended on the bounty of Nature, which on many small islands quickly ran out. In an early version of the California experience, the sea-shells left as remnants of ancient feasts became smaller and smaller as time went on, proof that people stripped the shores until they were almost bare. The hungry survivors perished or set out into emptiness. Many of the islands were wiped clean. Tonga's flightless birds and its iguanas were killed off within decades. The bones of turtles, seals and seabirds scattered around the abandoned homes of the colonisers show how soon the pillage started.

In a few places there appeared to be a period of grace before the collapse. On Easter Island the first Dutch explorers were astonished to find huge statues in its almost barren landscape. How had they been moved from the local quarries? Carbon dating suggested that people had arrived by around AD 600, on an island covered by palms and subtropical trees, while the statues were erected half a millennium later. Perhaps there had been a time of harmony before hubris set in. Now, alas, we know that men arrived later than supposed, around AD 1200, and set to graven images and vandalism almost at once. The forests, whose trees were cut down and used as rollers to move the stones, were destroyed within just a few years.

The damage has accelerated since the canoe gave way to the 747. The spread of the brown tree snake from New Guinea has been helped by its ability to hitch rides on aircraft. On many desert islands the reptile has killed off almost all the native birds and places once full of song are now still. A silent catastrophe is also under way beneath the waves. Ships fill up their ballast tanks in one tropical port and dump them in another. With the water comes a whole new set of migrants that – like the tree snakes – wipe out the locals.

As the Hawaiians discovered after the arrival of Captain Cook, aliens bring disease. Epidemics now rage beneath the seas. Where they came from is not known, for nobody was much interested in the health of the reefs until the plagues were well under way. Many such places have rotted from a dozen little-understood illnesses. In the 1970s and 80s the major shallow-water coral in the Caribbean and the Western Atlantic was almost wiped out by a progressive decay called white band disease. Nobody knows its cause, but the sinister black band disease and coralline lethal orange are caused by bacteria. They have both spread throughout the Pacific.

Exploitation became global in Victorian times. Thomas Henry Huxley himself, on a cruise on HMS *Rattlesnake* to the Great Barrier Reef, suggested that its dugongs be hunted for their oil. That idea came to naught, but soon a profitable *bêche-de-mer* (sea cucumber) fishery emerged. Huge sections were stripped for the Chinese trade, while European fishermen pillaged the turtles for their flesh and for the tortoiseshell used for fashionable hairbrushes in England.

Mother of pearl – nacre, the bright interior of marine shells such as abalones and pearl oysters – was also in demand for knife handles and the like. Within seven years of the discovery of the beds a hundred vessels were at work. As the shells became smaller the work got harder and the divers plunged deeper. Soon, Japanese boats dominated the trade. A pathetic endeavour to regulate matters achieved nothing and the fishery's failure – as inevitable as that of the white abalone – led to fury. The 1905 Australia Constitution Act makes reference to that ecological disaster: 'no coloured alien should be admitted for the pearl-shelling industry . . . steps should be taken to reduce this vast preponderance of aliens and to have this outpost guarded by a hardy population of loyal and patriotic Australians'. The ruin of the mother-of-pearl trade led in part to the White Australia Policy that lasted until after the Second World War.

On the Great Barrier Reef, public ownership and private profligacy went together. The pearl-fishers' patriotic outpost became the State of Queensland, a place that passed its history in a state of ecological banditry. Its reef was a prime target. In the 1960s oil exploration began and the premier, Sir Joh Bjelke-Petersen, sold leases to his own companies. Developers built concrete resorts on the most sensitive parts of the coast and those tropical Blackpools poured filth into the sea. Fantasy Island, a hotel on a concrete pontoon, was to be the first of many (it sank, and was renamed Fantasy Reef, but its owners were forced to tow the wreck to the mainland). The damage is impressive, for the population of sea-cows has dropped by 97 per cent since 1970. The creatures are now protected, but it may take a century for them to recover even to the diminished levels of those days.

Reefs have long been used for stone – the Old Town in Zanzibar is built of coral and very handsome it is – but many of the hideous concrete cities of the tropics are also based on the same stuff. A century ago the Queensland Cement and Lime Company began its operations in the seas off Brisbane. As a result, parts of that subtropical suburb are the cinders of burned corals. The city's yachtsmen, when they escape to the Barrier to drop anchor in some isolated spot, smash what lies below. Its merchant ships do the same.

Reefs that once fed those who live upon them have become a worldwide resource. Seafood caught in tropical seas travels for thousands of miles to satisfy a gourmet's tastes. In Hong Kong certain fish sell for more than £100 a kilo, while a plate of a delicacy based on the lips of the coral wrasse costs twice as much. The damage is sometimes reckless. A Taiwanese boat cut just a single muscle from each of fifteen thousand giant clams in a remote spot in the Western Caroline Islands before it was stopped. The clam population will take twenty years to recover, if it ever does.

Fishermen who once used spears or lines now prefer dynamite (or a plastic bottle filled with the IRA cocktail of nitrate fertiliser and diesel fuel). That catches lots of fish, but kills many more and blasts large holes in the coral itself. A boatload of dynamiters can explode a hundred bombs a day, enough to shatter an ecosystem. The habit did not become widespread until the 1980s, but in the Philippines has already destroyed three-quarters of the habitat. The fishermen are well aware of the dangers of their trade and many have had limbs blown off. Some now use cyanide instead, which kills the polyps just as well as it does their prey.

Some fish are worth more alive than dead. The fashion for seawater aquariums means that thirty million reef fish accompanied by several thousand tons of living rock are harvested each year, in a trade worth more than a hundred million dollars in the United States alone. Most of the material originates in the Philippines and Indonesia.

The global fish catch peaked in the late 1980s and has dropped each year since then. Nine-tenths of all large predators such as tuna and swordfish have been killed off. Now the hunters have moved on, to the deep seas, in huge ships subsidised by government as the common weal pays to destroy the common wealth. Fisheries a kilometre and more below the surface are aimed at the grenadier, the armour head, the Chilean sea bass and the orange roughy (rebaptised from its original name of 'slimehead'). Such animals live slow and careful lives. They can last for two centuries and may wait twenty years to breed. Many of the specimens on a fishmonger's slab are more than a hundred years old.

It took five centuries for fishermen to destroy the underwater gardens of the Mediterranean, two hundred years to pillage the Great Barrier Reef and twenty to kill the California abalone. Many of the cold deep-water corals discovered at the turn of the millennium will be gone within a decade.

At first the fishermen knew little (and cared less) about the ecology of their new-found prey, save that they were found in just a few remote places. In fact, many live on cold-water reefs or on the summits of sea-mounts. The hunt for the orange roughy can lift two tons of coral for each ton of fish. Parts of the Darwin Mounds have been flattened and it took Asian ships just a decade to clean up many of the Pacific sea-mounts. The United States Geological Survey now sells sonar maps of the bottom, top secret at the time of the Cold War, that make it possible to drop nets straight onto such places. On each sweep the gear smashes structures ten metres high and thousands of years old. In the distant oceans men fish not with spears or dynamite but with bulldozers, as heavy chains are followed by nets a hundred metres across and with rollers that can move a fifteen-ton boulder. The tragedy of the commons is seen at its starkest in the blackness of the deeps.

Fishermen know that their resources will not last, but they cannot afford to stop as rivals will step into the gap.

Private property came to Australia with Captain Cook and the sea soon paid the price. His first step ashore was on the Kurnell Peninsula in Botany Bay, then an open forest in which Aborigines hunted. The settlers cut down the trees and brought in sheep and cattle. Within a century most of the soil had washed into the bay and the Peninsula became an unstable sand-pile. The film *Mad Max: Beyond Thunderdome*, set in a post-nuclear desert, was made where the trees once stood.

In that forest, like all others, vegetation sprayed vapour into the heavens during daylight hours. It condensed at night. Woodland water is rather static. Some blows away as moisture, or flows off in streams, but most is recycled close to where it started, which means that not much earth is washed away. Fields and pastures, in contrast, are not closed systems for they often become sodden or parched and must be drained or irrigated. This has dire effects on the soil, which is swept off in rivers empowered by the loss of trees.

Lake Trummen in Sweden has had its bottom minutely examined. At the end of the Ice Age, with a shattered landscape around, it gained a millimetre of sediment a year. Within a few thousand years the figure dropped by nine-tenths as trees grew and stabilised the land. In the nineteenth century came a drastic change. The dose of mud increased a hundred times as the forests were cut down. Then, in the 1960s, the lake almost died. It was saved with a giant pump that spread the sludge over the surrounding landscape.

Sweden's problems are much magnified in the tropics. The muddy day of judgement for the Great Barrier Reef began as Europeans moved northwards. The tale is told in its native rock. Barium is rare in the ocean but common in river mud. The chemical is absorbed by the corals whenever the sea is hit by a surge of fresh water. A spike appeared soon after 1870 as the farmers began to destroy the rainforests. Australia still cuts down its trees faster than any other developed country; half of Queensland's native woodlands have gone and in some places erosion has gone up by ten times in a century. The nation's rivers pour out four times more nutriment than they did in the days before Captain Cook, and kill off plenty of corals as they do so.

In other places the damage is even worse. Seven-tenths of the reefs of the Philippines were choked in just a few decades. In the 1990s five hundred kilometres of the same habitat in Indonesia, not far away, dropped dead – an event without equal in history. A red tide, a bloom of poisonous plankton, was to blame. It was sparked off when airborne dust rich in iron, the nutrient that limits the yield of the far oceans, spewed into the air from giant fires set by farmers as they burned down jungles to make fields.

Sludge causes chaos in other ways. Ecologists speak of 'the paradox of enrichment', the fact that fertiliser increases the productivity of an ecosystem (which is why a field of wheat lays down carbon even faster than a reef) but reduces its diversity. Many of today's oceanic extinctions are due to excess. Populated coasts in the tropics generate a wall of microbial slime that creeps across the shallows and kills their inhabitants. Often, the sewage causes algal blooms that cut off the light and kill the polyps.

Other creatures see the slime as a free meal. The crown of thorns – a starfish half a metre and more across, with twenty or so arms covered in poisonous spines – has the interesting habit of everting its stomach over a patch of coral. It can kill whole sections of reef. For most of the time the crown of thorns is rare. Even so, a female can produce a hundred million eggs a year and now and again they explode in number. The first flare-up was noticed in 1962 on Green Island, off Cairns in Queensland. Divers removed tens of thousands of the animals and injected thousands more with formalin, to no apparent effect. Within a decade the starfish had marched three hundred kilometres to the south. In time it faded away – but in 1979 it was back, with millions on Green Island alone. The creature then sank back into obscurity, only to reach another peak at the turn of the century.

The finger was pointed at fishermen or climate change, but pollution is much involved. The crown of thorns is most abundant in places that receive lots of nutriment from local rivers and outbreaks tend to happen soon after heavy rains, when they are full of silt and slime. More silt means more starfish and the plague has, as a result, now spread to Fiji and Japan.

Filth comes in many flavours. Two hundred and fifty million gallons of oil leaked into the Persian Gulf after the 1991 war and plenty reached the local reefs. Industrial effluent is just as bad: some of the corals of Okinawa, much used as a source of calcium by those who hope to extend their lives, are themselves now dangerously contaminated with lead and mercury.

The chemistry of the oceans has altered even in places far from any factory. The change is sapping the foundations of every atoll. Carbon dioxide, when combined with water, is acid. Acid slows the rate at which reef-builders lay down skeletons and weakens their very being. In time, the rock itself is etched away. Today's surge in greenhouse gas means that within a century the seas will be more acidic than they have been for four hundred million years and their chemistry will move towards a new and volatile equilibrium. If the amount of carbon in the atmosphere doubles by the end of the century, as it may, the rate at which life can make stone could drop by half. More will then be lost from reefs than the polyps can lay down.

Whatever the long-term threat from carbon chemistry, the most immediate menace to the undersea world comes – as it did in the Permian – from climate change. After years of denial, almost everyone accepts that a great warming is upon us. When even the head of Shell Oil says that he sees no hope for humankind unless the problem is sorted out, it is time to worry.

People have always taken notice of the weather. Icelandic scribes wrote down the news about ice in 1500, the English showed interest in signs of spring soon afterwards and in France the date of the grape harvest became a matter of official concern. Proper records begin in the eighteenth century, when Anders Celsius came up with his scale. Since then the amount of information has exploded.

The climate has been unsettled since long before man took an interest, with ice age followed by mild spell over the past thousands and millions of years. Each time, the stonemasons and the seas that feed them returned it to equilibrium. Before the first reefs the planet sweltered, but over a billion years the builders cleaned the air of its insulator and allowed life to flourish. Changes in the balance of carbon salts in the sea led marine creatures to make massive (or, as today, more fragile) skeletons. When the masons were most active, greenhouse gas was sucked into their factories and the Earth entered a period of cold. In their more feeble moments the weather improved as the element found its way back into the sky.

On the scale of tens of millions of years we are now in rather a cool period, and the reefs are to blame. Fifty-five million years ago life was steamy indeed – a hardy swimmer could have braved the seas around the North Pole – but then the amount of carbon dioxide in the air began to drop as the corals did their work. The Arctic and Antarctic ice-sheets formed and have ebbed and flowed ever since.

Drills into the Antarctic ice track the tie of the fatal gas with climate over the past four hundred thousand years. The fit between the two is precise. Over a typical glacial cycle levels rise from around 180 parts per million to about three hundred (a figure well below that of today). A shift in carbon dioxide is followed almost at once by a change in temperature which, given the amount now present, is bad news indeed. In the past, the efforts of the ocean-dwellers have always brought it down but that remedy will not succeed soon enough to solve today's problem.

We have enjoyed a brief moment of relative warmth since the end of the last ice age a few thousand years ago. The longest of all such events – a thirty-thousand-year interglacial, four hundred thousand years ago – had levels of carbon dioxide similar to those just before the Industrial Revolution. Without interference we too might have been blessed with as many millennia of stability. Our imprudence has changed that, for it has put the climate cycle into reverse.

Almost never has the temperature altered as fast as in recent decades. Except for a couple of periods around 1530 and 1730, European winters since the Norman Conquest were colder than those of today. The worst of all was in 1708, when the Swedes lost their army to Russia in a September snowstorm. They have improved since then, a trend mitigated by a few Dickensian Christmases and a slight drop for thirty years after the end of the Second World War. The late twentieth century marked a dramatic rise in the thermometer – and the winters from 1973 onwards have been the mildest of all.

Summers have been less consistent: 1757 was torrid, but the atmosphere cooled down until the miserable year of 1902. The weather became a little better until the 1940s and then fell back for thirty years – but since 1977 change has been dramatic, with the fastest thermal shift ever seen in Europe. The problem will get worse. For two decades after 1960 Paris averaged a heatwave a year. By 2080 the number will double and each event will last for a week longer than now.

In the northern hemisphere the summer of 2003 was the warmest ever recorded and in Britain 2006 gained the accolade of the hottest twelve months since the invention of the thermometer. Nine of the ten sultriest years for which we have information have been in the past decade and a half. Most scientists agree that the last third of the twentieth century was the warmest period for a millennium (and even the sceptics, apart from those in the White House, accept that they beat the five-hundred-year record). South of the equator the situation is just as bad. Growth rings in Pacific corals, which wax and wane as the temperature rises and falls, prove that the rise has been continuous, with no trends towards cooler weather since the days of Christopher Columbus.

The oceans have the capacity to absorb far more energy than does the air. More than eight-tenths of the excess trapped since the 1950s has been soaked up and in many places the water down to a couple of hundred metres has increased in temperature by a fifth of a degree or so. The heat has penetrated further into the Atlantic, with its deep currents, than into the Pacific (which has less vertical movement) and has warmed the oceans of the southern hemisphere more than those of the north. Like that of a storage heater, the ocean's reservoir is bound to be released sooner or later, which means that its thermal sponge will not solve the long-term climate problem.

The prospect of a new Jurassic Park alarms people who feel at home in the temperate days of the twenty-first century. In 2004 the greenhouse effect was described by the British government's Chief Scientific Adviser as a bigger threat than terrorism. A year later the United Nations' International Panel on Climate Change projected a 'potentially devastating' increase. The glaciers on Africa's Mountains of the Moon – described by Plato as snow-covered, and safe in wintry glory since then – have halved in size in the past twenty years and will be lost within the next twenty. Soon levels of carbon dioxide will be close to those at the time when our planet had no ice at all.

Climate change makes useful fuel for gloom-mongers, but optimism tries to break in. Some human activities – the dust churned up by modern farms or the smoke from forest fires, for example – generate clouds that reflect sunlight back into space and may cool things down. Even so, most predictions have faced up to a global increase of around three or four degrees within the lifetime of a baby born today, with a conservative estimate of about half that and the most radical projecting as much as eleven degrees – which would mean catastrophe. The latest figures have been revised upwards after the unwelcome discovery that the disaster is feeding on itself, with more carbon dioxide released from land and sea as the planet heats up. If that effect is real, the problem may soon be out of control.

In England itself, the average temperature has gone up by one degree since the 1950s, and the nation's gardens, in climatic terms, are moving southwards each year at a rate equivalent to the length of a cricket pitch. Many European butterflies and birds have become common on this side of the Channel. Some species can migrate to escape, but many others do not have the chance. The thermal crisis may drive a quarter of all land plants and animals to extinction by the time the species most to blame reaches its peak of abundance in 2060.

In the oceans the situation is worse. The first symptom of damage was noticed twenty years ago. All over the Pacific polyps became pale shadows of their former selves. Some recovered, but many did not. Each attack of bleaching was the result of a breakdown in the relationship between the host and its symbiont. After a shock, the internal assistants leave and take their pigment with them. The other party loses its colour, its metabolism slows down, less protein is made and it may be obliged to give up sex. Certain kinds, such as the branched corals, are particularly susceptible to the problem.

In some species bleaching is a regular event, with a drop in symbionts each summer, but the latest attack is exceptional. Low tides, fresh water, infection or pollution can all spark off the malaise, but a hot spell is the most frequent culprit. An increase in water temperature of little more than a degree for just a week is enough to cause an outbreak. Some of the animals recover as the mercury drops, but others do not.

The phenomenon has been known since the nineteenth

century, but has much increased in the past two decades. On the Great Barrier Reef it was first noticed in 1980. Since then there have been four hundred attacks. In 1982 huge swathes of the Pacific lost their symbionts within a few months. In that El Niño year a heatwave struck and the reefs paled before it. It took the Galapagos a decade to recover. Soon afterwards, in 1998, another torrid episode killed or damaged a sixth of all the world's corals. The Maldives and Seychelles suffered near-ruin and within four years yet another onslaught scorched the atolls. A few of the places that lost their colour in those difficult summers now show signs of recovery, but the recent round has killed so many that they will take years to get back to normal, if they ever do.

Cores through the reefs show a series of heatwaves over the past few centuries. Many, no doubt, were accompanied by plagues of bleaching. Now the temperature has shot up at such speed that by the middle of this century such episodes will happen in most places in the tropics almost every year.

Given enough time the polyps and their allies might shift to a more comfortable place, as they often have in the past. Some of those at the edge of their present range might even appreciate a new and more tropical globe, but at the present rate they will not have time to do so. The reef-builders will drown before they can swim to safety.

As Londoners have noticed, the sea is on the rise. The Thames Barrier was built in the early 1980s and in that decade its gates were closed no more than once in a couple of years. Now the bascules roll upwards eight times as often, and a second, taller edifice will be needed within a couple of decades.

The global picture of sea-level change was until not long ago confused. Both land and water rise and fall, and in some places the land is ahead of the game as it rebounds from the burden of the last ice age. Just twenty or so marine stations, most of them in Europe and North America, had sea-level data over more than a century and while many showed a rise a few did not. Even so, a trawl through the mass of records between 1870 and 2004 shows a steady advance, moderate at first, then faster and, since 1950, at a global average of just under two millimetres a year.

Now we are better-informed. Since 1992 the satellite Poseidon, named after the Greek god of the sea, has scanned the surface of the ocean. Poseidon is accompanied, ten kilometres below and a minute ahead, by his planned successor Jason (of Argonaut fame). The mechanical deities fly more than a thousand kilometres above the surface and weave an endless web as they cross and re-cross their own tracks. They map the entire ocean every ten days. As the satellites race through the skies they send down radar pulses and measure the reflection from the surface. The delay between signal and echo gives the height of the sea to within a couple of centimetres – an accuracy equivalent to measuring the thickness of a penny on the pavement when viewed from a jet plane.

Poseidon revealed the presence of large mounds of water attracted by the gravitational force of sea-mounts, some with atolls perched on top. The satellite picks up waves and even ripples. More important, its radar pulses have found a gradual increase in the height of the sea at about two and a half millimetres a year in the past decade. That means fifteen centimetres by 2060 and twice that by the end of the century. The rise will, in part, be due to the inflow of fresh water, but around a quarter of the total is a result of the expansion of the ocean itself as the temperature goes up. As the heat percolates into the deeps this is bound to increase.

Sea-level changes have happened again and again in the past as the ice advances and retreats. Such shifts, with ice age followed by milder times and a matching surge or retreat of great waters, once appeared to be rather relaxed events. Cores into the Greenland and Antarctic ice-caps reveal a long cycle of temperature and of sea-level change over several hundred millennia, with each shift taking some twenty thousand years. Now and again there was a spurt of heat or cold and the oceans rose or fell, sometimes by many metres. They took – or so it seemed – a long time to respond to a shift in the thermometer. The reaction to a change in climate appeared to be so sluggish that corals should manage to keep up even with today's overheated world.

Unfortunately, the reefs themselves show that the polar record is blurred. Water takes time and energy to melt or to freeze. As a result, ice is slow to respond to climate change and gives an incomplete picture of any change in temperature. Polyps, in contrast, respond at once to shifts in sea level, for they die if exposed to fresh air or plunged into darkness. A drill through the rock reveals swings far more sudden than those hinted at by the Greenland ice-cap. Quite often the ocean rose by ten metres in ten centuries, which means that had the glaciers revolted against the invasion of William the Conqueror many cities in Britain would be submerged, segments of East Anglia would sink and most of the world's atolls would be lost.

Sudden collapses of the ice were to blame. Another crack-up is around the corner. The Arctic sea ice has shrunk by a quarter in the past thirty years, and in 2006, for the first time since modern humans evolved, a ship could have sailed in open water to the North Pole itself. The great Greenland glacier – that could, if it melted, raise the oceans by seven metres – is now shedding water as fast as the Nile and that rate seems to be accelerating. Nine-tenths of the Earth's ice is in Antarctica and the shelf around the continent has begun to break up, while the ice-sheets that feed it are now thawing at a great rate. Should all the polar ice be lost, north and south, the seas would surge by seventy metres and every reef on Earth would be engulfed.

Atolls such as Funafuti, the site of the Royal Society's experiment to test Darwin's theory, are already flooded at high spring tide. The island's drinking water has begun to turn salty and local politicians fear that they will be forced to abandon the place altogether. Not just humans will suffer. Several of the North-West Hawaiian Islands, home of millions of sea-birds and of unique monk-seals, are set to lose half their area within a century and there is rather desperate talk of dumping sand onto beaches to provide a refuge.

What is to be done about the reefs? Greenhouse gases have proved impossible to control and marine pollution has been almost as intractable. Some remedies verge on the desperate – fire sodium into the stratosphere to make high clouds to keep the sunlight out, place a filter in space between the Sun and the Earth, introduce compulsory vegetarianism to reduce carbon production from farms or even, with a certain cynicism, force people to wear white to reflect energy back into space.

Through the encircling gloom a few gleams of hope can be seen. The fiscal case is clear. Hawaii's coral fringes are estimated to generate $300 million a year in revenue while the Great Barrier Reef in Australia has been valued at thirty times as much. The United Nations estimates that, on the global scale, each square kilometre of reef is worth between a hundred thousand and a million dollars a year. Fish are valuable, tourists pay a lot to visit and the reefs' solid structures save vast sums as they protect the coasts that lie behind them.

Some islanders have realised how precious their homelands must be. Before Captain Cook, each Tahitian fiefdom stretched in a triangle from the island's peak down a valley, widening as it went until it reached the seaward edge of the rocky platform. Strict rules controlled who could fish, when, where and what for. On Yap, class did the job; poor people (and women) could fish in the tide-pools, the bourgeoisie were restricted to lines from shore while only aristocrats could take canoes to the outer barrier itself. Other islanders conserved their resources with taboos that could last for years, or with rules that calm places were forbidden except in stormy weather. As a result, marine life flourished and sustained vast populations. Not until the Western version of Utopia was introduced did the harvest collapse.

In a few places local management has returned. On Vanuatu a

hundred or so villages keep outsiders at bay and control their own sections of reef, while Samoa has banned the use of dynamite and poisons. Many coastal fisheries now look as if they may sustain themselves. The largest groupers – the fish that sell for high prices in Hong Kong – have been killed off in some places, but they are now farmed in parts of Asia, which reduces the pressure on natural populations. The giant clam, a delicacy in Japan, has already been lost from Fiji, Guam and other islands. In Palau, in contrast, the clams are protected. They are cultured rather than hunted, to give a valuable export.

A few of those fragile habitats have the good luck to belong to the rich. Australia defined its Great Barrier Reef Marine Park in 1975 and Tasmania now has a Sea-mount Marine Reserve to match. Then the Barrier was designated a World Heritage Site. At first the fishermen were kept out of a mere one part in twenty. Even that helped, for the protected sections soon held twice as many fish as before. In 2004 they were excluded from a third of the entire system. The Darwin Mounds were closed to trawlers in that year and soon afterwards such vessels were banished from a million square kilometres of coral gardens around the Aleutian Islands. In the United States a fifth of the nation's reefs are now protected. Hawaii has done a great deal to preserve its marine heritage. The live rock harvest was banned in the late 1990s, the sewage problem has been solved and the decline of the pineapple industry means that less silt pours into the sea. In 2006 came an absolute ban on fishing around the most distant islands in the chain.

The rage for conservation has spread and the globe now has around a thousand reef reserves that cover almost a fifth of the total. Not all the news is rosy. The main centres of diversity such as the East Indies Triangle, which covers the Philippines to Malaysia and New Guinea, and its equivalent in the Caribbean, belong to nations that cannot afford to support them. The region has preserved plenty of reefs, but most of the reserves allow fishermen in and are a magnet for poachers. Fewer than a hundred, a tiny fraction of the world total, are fully protected and most are too small to harbour the large fish that are their most valuable resource.

By accident or by design, man has sometimes given Nature a second chance. Many of the ships sunk in the Battle of the Coral Sea are now reefs in their own right and the Gulf of Mexico has three thousand new coral communities upon the legs of its oil platforms. More than a hundred redundant rigs have already been tipped over as foundations for new reefs. Artificial structures made of sunken ships, bags of sand or discarded tyres also play a part and are soon colonised by fish and other creatures. Half a million concrete reef-balls, with or without human ashes, now lie on the ocean floor. Such ersatz spots can be just as popular with tourists as are their natural equivalents. Dubai, that haunt of the super-rich, lacks an offshore barrier of its own but plans an artificial version to generate an attraction that Neptune himself has not been able to provide.

Other attempts to rescue the underwater world include controls on divers and on the craze for marine aquariums. In the Mediterranean the French have transplanted the red coral to artificial marine grottoes with some success and in the Indian Ocean there have been odd experiments in which electric currents applied to cages made of wire persuade polyps to grow with enormous speed.

In spite of all these efforts the world of coral gives more reason for despondency than for hope. Local conservation can do little in the face of global change. The future of the reefs is bleak indeed. Their end presages a catastrophe that will spread far beyond their bounds – and reminds us that we too are far from safe.

Captain Cook's tropical adventures came to their gruesome end on Hawaii's coral shores. They began in another supposed Island of the Blessed, Tahiti, where he measured the distance between the Earth and the Sun using the path followed by Venus as it passed in front of the solar disc.

The idea turned on the work of Isaac Newton who, seventy

years before, had come up with the notion of a universe ruled by gravity. The great scientist was as interested in prophecy as in physics. Like many of those who wring their hands about the state of the planet today, Newton was convinced of the imminence of the end. Just like them, he made some confident forecasts of when it would happen. Mathematics had uncovered the secrets of 'the book of God's works', the Universe, and must, he was sure, be able to do the same for 'the book of God's words', that 'history of things to come', the Bible. He penned three million words of learned Observations upon 'The Power of the Eleventh Horn of Daniel's fourth Beast, to change times and laws', 'Of the Kingdoms represented in Daniel by the Ram and the He-Goat' and so forth. Many of his works, heretical as they were, were kept secret.

At a certain point Newton calculates the date of Armageddon. He used a biblical phrase – time, times and a half (interpreted as 'a year, two years and a half', or forty-two months of thirty days each) – and turned those 1260 days into years. In AD 800 Charlemagne was crowned Emperor of Rome by the Pope: the moment at which, the physicist was sure, the Church fell into corruption. Add the two figures together and we get to the day of reckoning in the near future, in 2060, little more than half a century from now – and the very year in which the human population will reach its peak. For coral at least, Armageddon will by then be well under way. Many reefs have already been weighed in the balance and found wanting. By then many more will be dead.

Science and the end of all things still go together and with rather better evidence than that available to the author of 'The Ram and the He-Goat'. The reefs themselves make some firm – and sombre – predictions about our own apocalypse. Newton was right to worry about that disturbing event but was out in his timing. The polyps and their ancestors make a far more persuasive statement than does the Book of Revelation about just what the future holds.

Like trees, certain reef-builders have growth rings. They can be read on a daily, monthly and annual basis back to almost the origin of the first living creatures. Isaac Newton's birthday on Christmas Day comes at the end of a great darkness as the hours of daylight increase. The big picture looks bright, too, for the rhythms of growth seen in ancient reefs show that our planet's days have been getting longer since it was born. The maritime clock suggests that the timepiece upon which we live loses twenty seconds in each million years. The stromatolites, three billion years ago, saw the sun rise upon their work almost five hundred times each year, while the end-Permian extinction had 380 twenty-three-hour days in each annual cycle. My own memories of a distant Welsh childhood – like those of most sixty-year-olds – make those distant days feel endless, but the undersea evidence shows that each was in fact shorter than those I live through now. For all of us, young or old, they are set to stretch out even more.

Our planet is ruled by Newtonian mechanics; by gravity mitigated by friction. The Moon causes tides that grind the sea against the land. These in their turn slow down the Earth's rotation as they transfer energy to its satellite and cause the days to lengthen as they do so. The distance between the two bodies can be measured to within a millimetre with a laser beam bounced off the Moon's surface.

It increases by 3.8 centimetres each year, which means seventy-five metres – rather less than the length of a soccer pitch – since the Crucifixion. The process has gone on since our satellite was born. It explains both why today's corals see fewer sunrises than did their stromatolite predecessors and why modern youth can pack more into the period between sunrise and sunset than I ever could.

Physicists are now less concerned about apocalypse than was Newton, but his laws make predictions as inexorable as those of the Good Book itself. The days will continue to stretch out as surely as the tides rise and fall and in just fifteen billion years each will last a thousand hours. This seems reason to be cheerful but not all the portents are good. By that date the Moon will stand frozen in the sky, visible from just one side of the globe. Long before then, the reefs, their inhabitants and mankind itself will have nothing to worry about, for the Earth will be engulfed by a gigantic, red and moribund Sun.

The reefs tell the tale of how life began and record many of the catastrophes through which it has struggled. As human folly threatens their paradise with premature demise such places remind every one of us, pessimist or otherwise, that our own extinction is as certain as is theirs. Whether it will take place in the slow course of evolutionary time, or in the near future as our own imprudence causes Nature to take her revenge, neither Newton nor Darwin can say.

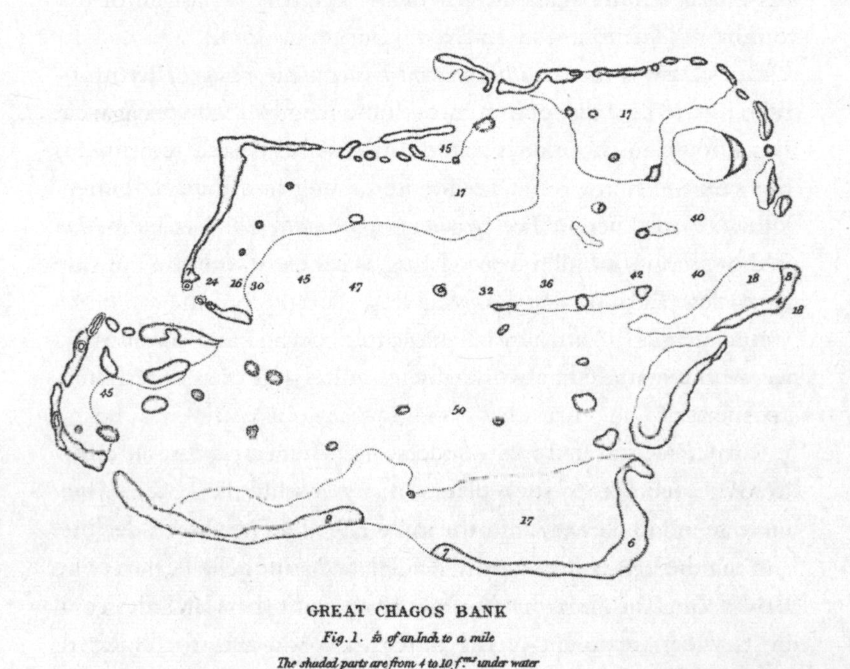

GREAT CHAGOS BANK

Fig. 1. $\frac{1}{10}$ of an Inch to a mile

The shaded parts are from 4 to 10 f.ms under water

LITERATURE CITED

Coral is made by an eclectic little bunch of animals that, as this book tries to show, illuminate topics as diverse as geology, genetics, evolution, ecology, development, meteorology and more. I am no authority on many (or perhaps any) of those sciences and have been obliged to read widely in subjects other than my own in the hope of obtaining at least a veneer of expertise.

The foundation stone of the science of corals was laid by Charles Darwin in *The Structure and Distribution of Coral Reefs*, published in 1842. His book goes to some lengths to describe what was known in those days and to discuss how each reef fits his theories and is as a result a less entertaining read than his better-known works such as *The Origin of Species* and *The Descent of Man* (although it is not such heavy going as his multi-volume opus on barnacles). Even so, *Coral Reefs* gives an insight into the mind of a young genius in an almost uncharted ocean, and among the *longueurs* inevitable in a work of science has moments of real literary merit.

Coral Reefs has had many successors. There is almost an infinity of guidebooks to such places, many lavishly illustrated. They increase in number at about the same rate as their subjects decline. Among the best is *Reef Life* by Andrea and Antonella Ferrari. The BBC's *The Blue Planet* by Andrew Byatt and others includes coral in its wider treatment of the undersea world and, for children, *Coral Reef: Around the Clock with the Animals of the Ocean* published by Dorling Kindersley is accurate and well produced.

More technical works include *The Life and Death of Coral Reefs* edited by Charles Birkeland; *The Great Barrier Reef: History, Science and Imagination* by James Bowen and Margarita Bowen and, by Mark D. Spalding and others, *World Atlas of Coral Reefs*. The great men (and they were almost all men) who contributed to the debate about their origins have been much biographised: Darwin, of course, in Janet Browne's *Charles Darwin*, but also Thomas Henry Huxley in Adrian Desmond's *T. H. Huxley: Evolution's High Priest* and their predecessor Captain Cook in the eccentric but engrossing *The Trial of the Cannibal Dog: The Remarkable Story of Captain Cook's Encounters in the South Seas* by Anne Salmond. The battles about the origin of reefs are well described in David Dobbs' *Reef Madness: Charles Darwin, Alexander Agassiz, and the Meaning of Coral*. Other books continue to appear, and those colonial and interlinked beings, the polyps, are well served by today's electronic networks.

The literature cited below represents a small part of the mass filtered for this book. I have concentrated on newer papers because the science is in a state of rapid advance. Earlier publications can be tracked down through more recent work.

Chapter I: The King of Cocos–Keeling

Allemand, D. and others 2004 'Biomineralisation in reef-building corals: from molecular mechanisms to environmental control', *Comptes Rendus Palevol* 3: 453–67.

Allwood, A. C. and others 2006 'Stromatolite reef from the Early Archaean era of Australia', *Nature* 441: 714–18.

Bruno, L. A. 2003 'The bequest of the nuclear battlefield: Science, nature and the atom during the first decade of the Cold War', *Historical Studies in the Physical and Biological Sciences* 33: 237–60.

Dalton, R. 2003 'A window on the inner Earth', *Nature* 421: 10–12.

Forges, B. R. D., Kolsow, J. A. and Poore, G. C. B. 2000 'Diversity and endemism of the benthic sea-mount fauna in the southwest Pacific', *Nature* 405: 944–7.

Henderson, G. M. 2005 'Coral clues to rapid sea-level change', *Science* 308: 361–2.

Hou, X. G. and others 2005 'Cambrian anemones with preserved soft tissue from the Chengjiang biota, China', *Lethaia* 38: 193–203.

Hua, Q. 2005 'Radiocarbon in corals from the Cocos (Keeling) Islands and implications for Indian Ocean circulation', *Geophysical Research Letters* 32 art. no. L21602.

Kunzig, R. 2000 *Mapping the Deep: the Extraordinary Story of Ocean Science* (Sort Of Books, London)

Malakoff, D. 2003 'Deep-sea mountaineering', *Science* 301: 1034–37.

Masson, D. G. and others 2003 'The origin of deep-water, coral-topped mounds in the northern Rockall Trough, Northeast Atlantic', *Marine Geology* 194: 159–80.

Montagionni, L. F. 2005 'History of Indo-Pacific coral reef systems since the last glaciation: Development patterns and controlling factors', *Earth Science Review* 71: 1–75.

Ohde, S. and others 2002 'The chronology of Funafuti Atoll: revisiting an old friend', *Proceedings of the Royal Society* A 458: 2289–306.

Parés, J. M. and Moore, T. C. 2005 'New evidence for the Hawaiian hotspot plume motion since the Eocene', *Earth and Planetary Science Letters* 237: 951–9.

Purdy E. G. and Winterer E. L. 2006 'Contradicting Barrier Reef relationships for Darwin's evolution of reef types', *International Journal of Earth Sciences* 95: 143–67.

Roberts, J. M. and others 2006 'Reefs of the deep: the biology and geology of cold-water coral ecosystems', *Science* 312: 543–7.

Stanley, G. D. 2003 'The evolution of modern corals and their early history', *Earth Science Review* 60: 195–225.

Thompson, W. G. and Goldstein, S. L. 2005 'Open-system coral ages reveal persistent suborbital sea-level cycles', *Science* 308: 401–4.

Vollmer, S. V. and Palumbi, S. R. 2002 'Hybridization and the evolution of coral reef diversity', *Science* 296: 2023–5.

Wood, R. 2001 'Biodiversity and the history of reefs', *Geological Journal* 36: 251–63.

Chapter II: The Hydra's Head

Baker, A. C. 2003 'Flexibility and specificity in coral-algal symbiosis: diversity, ecology and biogeography of *Symbiodinium*', *Annual Review of Ecology, Evolution and Systematics* 34: 661–89.

Ball, E. E. and others 2004 'A simple plan – cnidarians and the origins of developmental mechanisms', *Nature Reviews Genetics* 5: 567–77.

Broun, M. and others 2005 'Formation of the head organizer in *Hydra* involves the canonical *Wnt* pathway', *Development* 132: 2907–16.

Darling, J. A. and others 2005 'Rising starlet: the starlet sea anemone, *Nematostella vectensis*', *BioEssays* 27: 211–21.

David, C. N. and others 2005 '*Hydra* and the evolution of apoptosis', *Integrated Comparative Biology* 45: 631–8.

Galliot, B. and Schmid, V. 2002 'Cnidarians as model systems for understanding evolution and regeneration', *International Journal of Developmental Biology* 46: 39–48.

Holland, P. 2004 'The ups and downs of a sea-anemone', *Science* 304: 1255–6.

Kortschak, R. D. and others 2003 'EST analysis of the cnidarian *Acropora millepora* reveals extensive gene loss and rapid sequence divergence in the model invertebrates', *Current Biology* 13: 2190–5.

Kusserow, A. and others 2005 'Unexpected complexity of the *Wnt* gene family in a sea anemone', *Nature* 433: 156–60.

Lesser, M. P. 2004 'Experimental biology of coral reef ecosystems', *Journal of Experimental Marine Biology and Ecology* 300: 217–52.

Martindale, M. Q. 2005 'The evolution of metazoan axial properties', *Nature Reviews Genetics* 6: 917–27.

Morden, C. and Sherwood, A. R. 2002 'Continued evolutionary surprises among the dinoflagellates', *Proceedings of the National Academy of Sciences US* 99: 11558–60.

Muller, W. A., Teo, R. and Frank, U. 2004 'Totipotent migratory stem cells in a hydroid', *Developmental Biology* 275: 215–24.

Van Oppen, M. J. H. and others 2005 'The evolutionary history of the coral genus Acropora (Scleractinia, Cnidaria) based on a mitochondrial and a nuclear marker: reticulation, incomplete lineage sorting, or convergence?', *Molecular Biology and Evolution* 18: 1315–29.

Radtke, F. and Clevers, H. 2005 'Self-renewal and cancer of the gut: two sides of a coin', *Science* 307: 1904–9.

Reya, T. and Clevers, H. 2005 '*Wnt* signalling in stem cells and cancer', *Nature* 434: 843–50.

Seipp, S., Schmich, J. and Leitz, T. 2001 'Apoptosis – a death-inducing mechanism tightly linked with morphogenesis in *Hydractina echinata* (Cnidaria, Hydrozoa)', *Development* 128: 4891–8.

Sussmann, M. A. and Anversa, P. 2004 'Myocardial aging and senescence: where have the stem cells gone?', *Annual Review of Physiology* 66: 29–48.

Technau, U. and others 2005 'Maintenance of ancestral complexity and non-metazoan genes in two basal cnidarians', *Trends in Genetics* 21: 633–9.

Yamamato, M. and others 2005 'Regulation of oxidative stress by the anti-aging hormone Klotho', *Journal of Biological Chemistry* 280: 38029–34.

Chapter III: The Plover and the Crocodile

Arlotta, P. and Macklis, J. D. 2005 'Archeo-cell biology: Carbon dating is not just for pots and dinosaurs', *Cell* 122: 6–8.

Baker, A. C. and others 2004 'Coral reefs: Corals' adaptive response to climate change', *Nature* 430: 741–3.

Bekker, A. and others 2004 'Dating the rise of atmospheric oxygen', *Nature* 427: 117–20.

Bhattacharya, D., Yoon, H. S. and Hackett, J. D. 2004 'Photosynthetic eukaryotes unite: endosymbiosis connects the dots', *BioEssays* 26: 50–60.

Cavalier-Smith, T. 2003 'Genomic reduction and evolution of novel genetic membranes and protein-targeting machinery in eukaryote-eukaryote chimaeras (meta-algae)', *Philosophical Transactions of the Royal Society* B 358: 109–34.

Coffroth, M. A. and Santos, S. R. 2005 'Genetic diversity of symbiotic dinoflagellates in the genus *Symbiodinium*', *Protist* 156: 19–34.

Doolittle, W. F. and others 2003 'How big is the iceberg of which organellar genes in nuclear genomes are but the tip?', *Philosophical Transactions of the Royal Society* B 358: 39–58.

Dyall, S. D., Brown, M. T. and Johnson, P. J. 2004 'Ancient invasions: from endosymbionts to organelles', *Science* 304: 253–7.

Furla, P. and others 2005 'The symbiotic anthozoan: A physiological chimera between alga and animal', *Integrative and Comparative Biology* 45: 595–604.

Habetha, M. and others 2003 'The *Hydra viridis/Chlorella* symbiosis. Growth and sexual differentiation in polyps without symbionts', *Zoology* 106: 101–8.

Hay, M. E. and others 2004 'Mutualisms and aquatic community structure: The enemy of my enemy is my friend', *Annual Review of Ecology, Evolution and Systematics* 35: 175–97.

Lane, N. 2006 'Mitochondrial disease', *Nature* 440: 600–2.

Loeb, L. A. and others 2005 'The mitochondrial theory of aging and its relationship to reactive oxygen species damage and somatic mtDNA mutations', *Proceedings of the National Academy of Sciences US* 102: 18769–70.

Muscatine, L. and others 2005 'Stable isotopes of organic matrix from coral skeleton', *Proceedings of the National Academy of Sciences US* 102: 1525–30.

Pochon, X. and others 2006 'Molecular phylogeny, evolutionary rates, and divergence timing of the symbiotic dinoflagellate genus *Symbiodinium*', *Molecular Phylogenetics and Evolution* 38: 20–30.

Rivera, M. C. and Lake, J. A. 2004 'The ring of life provides evidence for a genome fusion origin of eukaryotes', *Nature* 431: 152–5.

Rodriguez-Lanetty, M. and others 2006 'Transcriptome analysis of a cnidarian – dinoflagellate mutualism reveals complex modulation of host gene expression', *Genomics* 7: 23–9.

Ryan, F. 2002 *Darwin's Blind Spot: Evolution beyond Natural Selection* (Houghton Mifflin, New York).

Sachs, J. L. and Wilcox, T. P. 2006 'A shift to parasitism in the jellyfish symbiont, *Symbiodinium microadriaticum*', *Proceedings of the Royal Society* B 273: 425–9.

Stanley, G. D. 2006 'Photosymbiosis and the evolution of modern coral reefs', *Science* 312: 857–8.

Temple, M. D., Perrone, G. G. and Dawes, I. W. 2005 'Complex cellular responses to reactive oxygen species', *Trends in Cell Biology* 15: 319–27.

Wallace, D. 2005 'A mitochondrial paradigm of metabolic and degenerative diseases, aging and cancer: a dawn for evolutionary medicine', *Annual Review of Genetics* 39: 359–407.

Yuan, X. and others 2005 'Lichen-like symbiosis 600 million years ago', *Science* 308: 1017–20.

Chapter IV: The Empire of Chaos

Anderson, A. 2001 'No meat on that beautiful shore: the prehistoric abandonment of subtropical Polynesian islands', *International Journal of Osteoarchaeology* 11: 14–23.

Bellwood, D. R. and Hughes, T. P. 2001 'Regional-scale assembly rules and biodiversity of coral reefs', *Science* 292: 1532–4.

Bellwood, D. R. and others 2006 'Functional versatility supports coral reef biodiversity', *Proceedings of the Royal Society of London* B 273: 101–7.

Buckley, H. R. and Tayles, N. 2003 'Skeletal pathology in a prehistoric Pacific Island sample: Issues in lesion recording, quantification, and interpretation', *American Journal of Physical Anthropology* 122: 303–24.

Cane, M. 2005 'The evolution of El Niño, past and future', *Earth and Planetary Science Letters* 230: 227–40.

Chave, J. 2004 'Neutral theory and community ecology', *Ecology Letters* 7: 241–53.

Gagan, M. K. and others 2004 'Post-glacial evolution of the Indo-Pacific warm pool and El Niño-Southern Oscillation', *Quaternary International* 118: 127–43.

Gedalof, Z., Mantua N. J. and Peterson D. L. 2002 'A multi-century perspective of variability in the Pacific Decadal Oscillation: new insights from tree rings and coral', *Geophysical Research Letters* 29: 10–19.

Grottoli, A. G. and others 2003 'Decadal timescale shift in the ^{14}C records of a central equatorial Pacific coral', *Radiocarbon* 45: 91–9.

Hughes, T. P. and others 2005 'New paradigms for supporting the resilience of marine ecosystems', *Trends in Ecology and Evolution* 20: 380–6.

Hurles, M. E. and others 2003 'Native American Y chromosomes in Polynesia; the genetic impact of the Polynesian slave trade', *American Journal of Human Genetics* 72: 1282–7.

Karlson, R. H., Cornell, H. V. and Hughes, T. P. 2004 'Coral communities are regionally enriched along an oceanic biodiversity gradient', *Nature* 429: 867–9.

Lesser, M. P. 2004 'Experimental biology of coral reef ecosystems', *Journal of Experimental Marine Biology and Ecology* 300: 217–52.

Linsley, B. K. and others 2004 'Geochemical evidence from corals for changes in the amplitude and spatial pattern of South Pacific interdecadal climate variability over the last 300 years', *Climate Dynamics* 22: 1–11.

Lockwood, J. G. 2001 'Abrupt and sudden climatic transitions and fluctuations: a review', *International Journal of Climatology* 21: 1153–79.

Ostrander, G. K. and others 2000 'Rapid transition in the structure of a coral reef community: the effects of coral bleaching and physical disturbance', *Proceedings of the National Academy of Sciences US* 97: 5297–302.

Trenberth, K. E. and Otto-Bliesner B. L. 2003 'Toward integrated reconstruction of past climates', *Science* 300: 589–91.

Vigny, C. and others 2005 'Insight into the 2004 Sumatra–Andaman earthquake from GPS measurements in southeast Asia', *Nature* 436: 201–6.

Chapter V: The Maharajah's Jewels

Adams, J. M. and Piovesan, G. 2002 'Uncertainties in the role of land vegetation in the carbon cycle', *Chemosphere* 49: 805–19.

Archer, D. 2005 'Fate of fossil fuel CO_2 in geologic time', *Journal of Geophysical Research – Oceans*, 110 art. no. C09S05.

Barker, S., Higgins, J. A. and Elderfield, H. 2003 'The future of the carbon cycle: calcification response, ballast and feedback on atmospheric CO_2', *Philosophical Transactions of the Royal Society* A 361: 1977–99.

Bigg, G. R. and others 2003 'The role of the oceans in climate', *International Journal of Climatology* 23: 1127–59.

Canfield, D. E. 2005 'The early history of atmospheric oxygen: homage to Robert M. Garrels', *Annual Review of Earth and Planetary Sciences* 33: 1–36.

Coale, K. H. and others 2004 'Southern Ocean iron enrichment experiment: carbon cycling in high- and low-Si waters', *Science* 304: 408–14.

Hegerl, G. C. and others 2006 'Climate sensitivity constrained by temperature reconstructions over the past seven centuries', *Nature* 440: 1029–32.

Lal, R. 2004 'Soil carbon sequestration impacts on global climate change and food security', *Science* 304: 1623–7.

Montanez, I. P. 2002 'Biological skeletal carbonate records changes in major-ion chemistry of paleo-oceans', *Proceedings of the National Academy of Sciences US* 99: 15852–4.

Orr, J. C. and others 2005 'Anthropogenic ocean acidification over the twenty-first century and its impact on calcifying organisms', *Nature* 437: 681–6.

Ridgwell, A. and Zeebe, R. B. 2005 'The role of the global carbonate cycle in the regulation and evolution of the Earth system', *Earth and Planetary Science Letters* 234: 299–315.

Service, R. F. 2004 'The carbon conundrum', *Science* 305: 962–3.

Takahashi, T. 2004 'The fate of industrial carbon dioxide', *Science* 305: 352–3.

Zimov, S. A. and others, 2006 'Permafrost and the global carbon budget', *Science* 312: 1612–13.

Envoi: A Pessimist in Paradise

Bellwood, D. R. and others 2004 'Confronting the coral reef crisis', *Nature* 429: 827–33.

Benton, M. J. and Twitchett, R. J. 2003 'How to kill (almost) all life: the end-Permian extinction event', *Trends in Ecology and Evolution* 18: 358–65.

Briggs, J. C. 2005 'Coral reefs: conserving the evolutionary sources', *Biological Conservation* 126: 297–305.

Brodie, J. and others 2005 'Are increased nutrient inputs responsible for more outbreaks of crown-of-thorns starfish? An appraisal of the evidence', *Marine Pollution Bulletin* 51: 266–78.

Douglas, A. E. 2003 'Coral bleaching – how and why?', *Marine Pollution Bulletin* 46: 385–92.

Du Toit, J. T. and others 2004 'Conserving tropical nature; current challenges for ecologists', *Trends in Ecology and Evolution* 19: 12–17.

Garrabou, J. and Harmelin, J. G. 2002 'A 20-year study on life-history traits of a harvested long-lived temperate coral in the NW Mediterranean: insights into conservation and management needs', *Journal of Animal Ecology*, 71: 966–78.

Graham, A. A. J. and others 2006 'Dynamic fragility of oceanic coral reef ecosystems', *Proceedings of the National Academy of Sciences US* 103: 8425–9.

Hansen, J. and others 2005 'Earth's energy imbalance: confirmation and implications', *Science* 308: 1431–5.

Hasselmann, K. and others 2003 'The challenge of long-term climate change', *Science* 302: 1923–5.

Hegerl, G. C. and Bindoff, N. L. 2005 'Warming the world's oceans', *Science* 309: 254–5.

Hoegh-Guldberg, O. 2005 'Low coral cover in a high-CO_2 world', *Journal of Geophysical Research* 110 art. no. C09S06.

Hughes, T. P. and others 2003 'Climate change, human impacts and the resilience of coral reefs', *Science* 301: 929–33.

Jackson, J. B. C. 2001 'What was natural in the coastal oceans?', *Proceedings of the National Academy of Sciences US* 98: 5411–18.

Johannes, R. E. 2002 'The renaissance of community-based marine resource management in Oceania', *Annual Review of Ecology, Evolution and Systemics* 33: 317–40.

Knowlton, N. 2001 'The future of coral reefs', *Proceedings of the National Academcy of Sciences US* 98: 5419–25.

Lynas, M. 2004 *High Tide: News from a Warming World* (Flamingo, London)

McWilliams, J. P. and others 2005 'Accelerating impacts of temperature-induced coral bleaching in the Caribbean', *Ecology* 86: 2055–60.

Mora, C. and others 2006 'Coral reefs and the global network of marine protected areas', *Science* 312: 1750–3.

Pandolfi, J. M. and others 2003 'Global trajectories of the long-term decline of coral reef ecosystems', *Science* 301: 955–7.

Payne, J. L. and others 2004 'Large perturbations of the carbon cycle during recovery from the end-Permian extinction', *Science* 305: 506–9.

Pomeroy, R. S. and others 2006 'Farming the reef: is aquaculture a solution for reducing fishing pressure on coral reefs?', *Marine Policy* 30: 111–30.

Rampino, M. R. and Caldeiro, C. 2005 'Major perturbation of ocean chemistry and a "Strangelove Ocean" after the end-Permian mass extinction', *Terra Nova* 17: 554–9.

INDEX

THE SINGLE HELIX

A Turn Around the World of Science

Steve Jones

'Jones once again shows that, for all its difficulties,
science can still be fun'
Independent

The Single Helix is a miscellany of a hundred easy pieces about science. It brings to life a vast range of subjects united under the banner of scientific truth – the universal solvent that brings clarity to almost all the mysteries of the world we live in.

From chaos in the heavens to the fight against creationism, from optical illusions in tartan to the mathematics of elections and what rules the sex lives of cats, *The Single Helix* is a scientist's look at sciences other than his own – and as a result its author has been forced to make the complicated simple enough for even a biologist to understand.

'If anyone is capable of giving science a makeover it is Professor Steve Jones. And to judge by his collection of essays on science and its practitioners, he is as much Jack Dee and Ph.D. In the process he has produced one of the most engaging and revealing portraits of science and its practitioners you'll ever read'
Daily Express

'In his writing, as well as his manner, Jones is the
Alan Bennett of science'
Financial Times

Abacus
978 0 349 11940 3

Y: THE DESCENT OF MAN

Steve Jones

'Delightful, witty, insightful . . . a delicious romp through the biology of the human male'
Robin McKie, Observer

In his award-winning *Almost Like A Whale*, acclaimed science writer Steve Jones updated the book of the millennium: Charles Darwin's *The Origin of Species*. Now, taking his cue from Darwin's second great work, *The Descent of Man*, he turns his attention to one of nature's most neglected creatures: men.

'Steve Jones is the complete bollocks'
Bob Geldof

'Jones is much harder on men than I am'
Germaine Greer

'*Y:The Descent of Man* presents the male sex as a threatened species clinging to the fringes of life by virtue of its frail Y chromosome. Seldom has such a gloomy prediction been offered with so much zest'
Doris Lessing, *Sunday Telegraph*

'Jones' forte is genetics and he has an unrivalled ability to bring that important but often unintelligible science to life . . . he is an expert but somehow makes the reader feel like an equal'
Mark Ridley, *Sunday Times*

'Science communication at its best: up to date, authoritative, witty and packed with human interest . . . and not just a book for blokes: required reading, too for every woman who wants to know her enemy'
New Scientist

Abacus
978 0 349 11389 0